Selected Papers from the 3rd Ellisras Longitudinal Study and Other Non-communicable Diseases Studies International Conference

Selected Papers from the 3rd Ellisras Longitudinal Study and Other Non-communicable Diseases Studies International Conference

Guest Editor

Kotsedi Daniel Monyeki

MDPI

Basel • Beijing • Wuhan • Barcelona • Belgrade • Novi Sad • Cluj • Manchester

Guest Editor
Kotsedi Daniel Monyeki
Department of Physiology
and Environmental Health
University of Limpopo
Sovenga
South Africa

Editorial Office
MDPI AG
Grosspeteranlage 5
4052 Basel, Switzerland

This is a reprint of the Special Issue, published open access by the journal *Children* (ISSN 2227-9067), freely accessible at: https://www.mdpi.com/journal/children/special_issues/3rd_Ellisras_ncd_study.

For citation purposes, cite each article independently as indicated on the article page online and as indicated below:

Lastname, A.A.; Lastname, B.B. Article Title. *Journal Name* **Year**, *Volume Number*, Page Range.

ISBN 978-3-7258-6838-4 (Hbk)
ISBN 978-3-7258-6839-1 (PDF)
https://doi.org/10.3390/books978-3-7258-6839-1

Contents

About the Editor

Kotsedi Daniel Monyeki

Kotsedi Daniel Monyeki, a full professor at the University of Limpopo in the Physiology and Environmental Health Department, obtained his PhD from Potchefstroom University for CHE and the MPH University of Cape Town, South Africa. He is a C2-rated researcher by the South Africa National Research Foundation. He is an Editorial Board Member of the *Journal of Human Hypertension*, UK, and *Children*, MDPI. He founded the PHASHASHA Newsletter, which is based on major findings regarding non-communicable diseases in other studies, including the ELS, which is distributed to the community in hard copy format during the Ellisras Longitudinal Study and Other Non-Communicable Diseases Studies International Conferences. The language used in the PHASHASHA Newsletter can be understood by ordinary people in the community, helping them to make healthy choices regarding their lifestyle. Dan recently secured funding from the European Commission under the auspices of Horizon 2020 as a co-investigator on the project "Scaling up Packages of Interventions for Cardiovascular Disease Prevention in Selected Sites in Europe and Sub-Saharan Africa: An Implementation Research" (SPICES Project). He is the principal investigator in the Ellisras Longitudinal Study funded by the South African National Research Foundation, the South African Medical Research Council, the University of Limpopo, and the Vu Medical University Centre Amsterdam.

Preface

The profile of non-communicable diseases (NCDs) is changing rapidly amongst rural African populations due to COVID-19, poverty, and other psychosocial factors. Unfortunately, low literacy levels impact the triple burden of disease, and the prevalence of recommended lifestyle changes remains extremely low among individuals, families, and communities. Health promotion initiatives by medical experts, academics, and scholars are crucial if one wants to be successful in changing an individual's lifestyle. The community must look for novel and creative solutions to address demographic challenges like overweight/obesity, hypertension, physical inactivity, smoking cessation, and alcohol abuse in order to maintain a healthy lifestyle, Furthermore, to work towards alleviating the burden of NCDs today, it is crucial to have a thorough understanding of risk factors for CVD, such as behavioral, clinical, and sub-clinical, as well as the resulting target organ damage. To succeed in changing the lifestyle of an individual, the first requirement is the eradication of illiteracy by health professionals, academics, and scholars by providing primary health information to these sectors of the community through personal interactions.

Health professionals, academics, scholars, and community members from all parts of the world gathered at the University of Limpopo and Kitty Village, Lephalale, South Africa between 22 and 24 November 2022 under the unique context of the 3rd Ellisras Longitudinal Study and Other Non-Communicable Diseases Studies International Conference (ELSONCDSIC). The theme of the conference was "Multi-morbidity, poverty and COVID 19: Are we winning?" The purpose of the 3rd ELSONCDIC was to provide a unique, inclusive platform for discussion by ordinary members of the Ellisras community, expert scholars, students, and experienced professionals from all over the world to share valuable knowledge on NCDs, such that the conference participants and Ellisras/Lephalale communities were able to take back the knowledge gained to their respective communities and families.

Kotsedi Daniel Monyeki
Guest Editor

children

MDPI

Article

Assessment of the Performance of Obesity Measures in the Screening of Hypertension in a South African Adolescent Population

Godwill Azeh Engwa [1,*], Paul Chungag Anye [2], Nandu Goswami [1,3,4,5] and Benedicta Ngwenchi Nkeh-Chungag [1]

1 Department of Biological and Environmental Sciences, Faculty of Natural Sciences, Walter Sisulu University PBX1, Mthatha 5117, South Africa; nandu.goswami@medunigraz.at (N.G.); bnkehchungag@wsu.ac.za (B.N.N.-C.)
2 MBCHB Programme, Faculty of Health Sciences, Walter Sisulu University PBX1, Mthatha 5117, South Africa; 214023362@wsu.ac.za
3 Gravitational Physiology and Medicine Research Unit, Division of Physiology and Pathophysiology, Otto Loewi Research Center of Vascular Biology, Immunity and Inflammation, Medical University of Graz, 8036 Graz, Austria
4 Department of Health Sciences, Alma Mater Europaea, 2000 Maribor, Slovenia
5 College of Medicine, Mohammed Bin Rashid University of Medicine and Health Sciences, Dubai P.O. Box 505055, United Arab Emirates
* Correspondence: gengwa@wsu.ac.za

Abstract: Obesity is known to be one of the most significant risk factors for essential hypertension in childhood. However, whether obesity cut-offs may predict hypertension screening in adolescents remains controversial. This study investigated the performance of obesity cut-off values for the screening of hypertension in a South African adolescent population. In this cross-sectional study, 1144 adolescents aged between 11 and 17 years were recruited from the Eastern Cape Province of South Africa. Anthropometric and blood pressure (BP) parameters including diastolic blood pressure (DBP), systolic blood pressure (SBP), and heart rate (HR) were measured. Assessment of the sensitivity and specific of obesity cut-off values in predicting hypertension was performed using receiver operating characteristic (ROC) analysis. Obesity was positively associated ($p < 0.01$) with hypertension. Obese individuals, based on the predicted obesity using BMI percentile cut-off (pBMI85.2%), were more likely to develop hypertension (odds ratio: 2.070; $p < 0.001$) than their counterparts based on the observed obesity cut-off (pBMI95%) (Odd ratio: 1.748 $p = 0.004$). The area under the curve (AUC) of BMI percentile and WHtR for screening SBP percentile, and DBP percentile and HR as per ROC analysis, was low (<0.65). Equally, the sensitivity and specificity were low (<0.6) for all BP measures (SBP, DBP, and HR). Furthermore, the cut-off values for blood pressure measures, as established by ROC analysis using anthropometric measures, were far below the recommended cut-off values for hypertension screening. The obesity cut-offs for BMI percentile and WHtR established in this populations showed poor performance in diagnosing hypertension even though they were strong predictors of hypertension.

Keywords: obesity; hypertension; adolescents; cut-off values; diagnostic performance

1. Introduction

Hypertension continues to be a global problem as the predominant cause of cardiovascular and metabolic-related diseases and a major cause for premature death worldwide [1]. Although hypertension is a commonly observed disease in adults, it is becoming a major concern in young adults and even children. Childhood essential hypertension is a health problem of concern as it can progress to hypertension in adulthood [2,3]. This concern is attributed to the increased hypertension prevalence in children as well as adolescents [4].

According to a systematic review and meta-analysis in 2018, the global pool prevalence of childhood hypertension is about 4% and approximately 9% for prehypertension [5]; meanwhile, in Africa, a systematic review and meta-analysis in 2022 reported a pooled prevalence of 7·45% and 11% for hypertension and elevated blood pressure, respectively, in children [6]. A much higher prevalence has been reported in South Africa, with a systematic review reporting a range from 7.5% to 22.3% for hypertension prevalence in South African children [7]. Other studies in the Eastern cape province of South Africa have shown an even higher prevalence of hypertension in children [8] and adolescents [9].

Although not the sole contributor to the development of hypertension, obesity is known as one of the major predictors of hypertension [10]. Previous studies have reported that obese and overweight children have a higher risk for hypertension compared to non-obese children [11,12]. Various anthropometric indices have been used to screen obesity including measures of general obesity such as mid upper-arm circumference (MUAC), neck circumference (NC), and body mass index (BMI), as well as measures of central obesity such as hip circumference (HC), waist circumference (WC), and their ratios, waist–height ratio (WHtR) and waist–hip ratio (WHR) [13]. There is, however, no consensus on which anthropometric measures are better predictors of childhood hypertension, as certain studies claim measures of central obesity (WHR, WC and WHtR) [14,15] to be superior to general obesity (BMI) [16–18] while in other studies, it is not distinct [19]. Moreover, screening hypertension in adolescents and children is more complicated than in the adult population as childhood hypertension is dependent on height, gender, and age-specific references in diastolic blood pressure (DBP) and systolic blood pressure (SBP) [20].

In our previous study [21], we established new cut-off values for obesity screening in adolescents in a South African population. Our findings, based on ROC analysis, showed that BMI percentile (pBMI%) and WHtR were the most suitable measures for obesity screening in adolescents. With the new cut-offs for pBMI% and WtHR, the predicted prevalence of obesity increased compared to the observed prevalence based on the WHO [22] and the CDC [23] cut-off references. However, it remains unclear how these new cut-offs will predict hypertension in this population. Hence, the aim of this study was to assess the relationship between blood pressure and obesity based on these new cut-off values and assess their performance to predict hypertension in adolescents in a South African population.

2. Materials and Methods

2.1. Study Design and Population

A cross-sectional study which recruited 1144 South African adolescents including 348 males and 796 females aged between 11 and 17 years old was conducted. The participants were recruited from Mthatha, Libode, Alice, and East London of the Eastern Cape Province inhabited by South Africans of African origin.

2.2. Ethical Consideration

The study was approved by the Ethics Committee of University of Fort Hare (CH1011SCHU01) on 1 June 2016, and the Faculty of Health Sciences, Walter Sisulu University (Ref No: 112/2018), on 31 October 2018, South Africa. Prior to enrolment into the study, written informed consent was obtained from the parents or legal guardians of the children. Participants' data was kept confidential in accordance with the South Africa National Data Protection Act.

2.3. Blood Pressure Measurements

After resting for 5 min in a quiet room while sitting, participants were fitted with arm-size appropriate cuffs and blood pressure was measured three times at every 2 min interval using an automated sphygmomanometer (HBP-1100, Omron Healthcare Co., Kyoto, Japan). The mean blood pressure (BP) was calculated and then age, sex, and height were used to convert the BP to percentiles according to the guidelines for children (Centre for Disease Control and Prevention-National Health and Nutrition Examination Survey) [24]. Participants were classified as normotensive (NT) when diastolic BP (DBP) and systolic BP (SBP) was <90th percentile; pre-hypertensive (preHT) when DBP and SBP > 90th < 95th percentile; or hypertensive (HT) when DBP and/or SBP $\geq$ 95th percentile. Mean arterial pressure (MAP) was calculated as per the formula: MAP = (SBP + (2 $\times$ DBP))/3.

2.4. Anthropometric Measurements

Anthropometric measurements were taken in accordance with the International Standards for Anthropometric Assessments [25]. A wall-mounted Harpenden stadiometer was used to measure height in meters (m). Tanita weight scale (BC1000, Tanita Corporation, Tokyo, Japan) was used to measure the weight (kg), and the height was entered in the device to calculate the body mass index (BMI) expressed as weight/height^2 (kg/m^2). BMI, taking into consideration the age, sex and height of the children, was converted to BMI percentiles (pBMI) and classified as underweight: <5th percentile; normal weight: $\geq$5th to <85th percentile, overweight: $\geq$85th to <95th percentile; and obese: $\geq$95th percentile [26]. An anthropometric tape was used to measure the waist circumference (WC), hip circumference (HC), mid-upper arm circumference (MUAC), and neck circumference (NC) in centimetres (cm). WC and height were used to calculate the waist-to-height ratio (WHtR) and obesity was considered above a cut-off value of 0.5, as previously reported [27].

2.5. Statistical Analysis

The prevalence of hypertension was calculated as (presence of hypertension/total population) $\times$ 100. Predicted obesity was defined as obesity using the new cut-off values established in our previous study [21], while observed obesity is obesity based on the WHO and the CDC reference cut-off values. Statistical Package for Social Sciences (SPSS) software (version 20, IBM SPSS Inc., 2011, Chicago, IL, USA) was used to analyse data. Data were presented as mean $\pm$ standard deviation in tables. The mean differences between groups were analysed by independent sample t-test after adjusting for sex and age. Chi-square test was used to assess the association of obesity and hypertension by comparing categorical variables between groups while binary logistic regression was employed to determine the odds ratio of obesity to predict hypertension. Receiver operating characteristic (ROC) curve analysis [28] was used to determine the performance of anthropometric measurements to predict blood pressure and the area under the curve (AUC), an indicator to measure how precise an anthropometric measure distinguishes hypertension. The AUC ranged between 0 and 1. A value of 0.5, indicated by a diagonal line, shows that the anthropometric measure has no predictive performance for hypertension while a value of 1 indicates an ideal or optimal performance. The accuracy of anthropometric measures in predicting hypertension was determined by the specificity and was obtained from the ROC curve. The ROC curve is a plot of the sensitivity (true-positive rate) against 1-specificity (false-positive rate) for each anthropometric measure. The Youden index (value of the largest sum of sensitivity and specificity-1) was used to determine the optimal cut-off value for each anthropometric index (HC, WC, BMI, WHtR, and MUAC). A difference was considered to be significant at $p \leq 0.05$.

3. Results

3.1. Characteristics of Study Participants

A total of 193 adolescents out of 1144 were hypertensive giving a prevalence of 16.9%. Among females, 17.1% (136) were hypertensive while 16.4% (57) of males were hypertensive. The population was matched for weight between boys and girls, although boys were slightly taller than girls. WC, BMI, pBMI%, and WHtR were similar between boys and girls ($p > 0.05$) except for HC, NC, and MUAC which were higher in girls than boys ($p < 0.05$). All anthropometric measures were higher in hypertensive participants compared to non-hypertensive participants ($p < 0.05$). Blood pressure measures were higher in hypertensive participants ($p < 0.01$), as well as SBP and HR which were higher in females than males ($p < 0.05$) (Table 1).

Table 1. Baseline characteristics between hypertensive and non-hypertensive.

	Female		Male		p-Value		
	HT (136)	**NT (660)**	**HT (57)**	**NT (291)**	**HT**	**Sex**	**HT × Sex**
Weight (kg)	58.42 ± 14.13	54.75 ± 15.23	58.84 ± 16.29	49.43 ± 14.66	<0.001	0.283	0.211
Height (m)	1.56 ± 0.07	1.57 ± 0.08	1.61 ± 0.11	1.55 ± 0.11	0.019	0.723	0.043
WC (cm)	74.99 ± 10.19	72.52 ± 11.24	74.93 ± 11.34	69.79 ± 10.47	<0.001	0.825	0.341
HC (cm)	95.88 ± 12.42	91.26 ± 13.25	91.21 ± 13.69	86.23 ± 11.33	<0.001	0.036	0.547
NC (cm)	30.21 ± 1.98	29.84 ± 2.38	30.71 ± 3.45	29.87 ± 2.41	0.021	0.345	0.327
MUAC (cm)	25.29 ± 5.01	24.39 ± 4.31	24.51 ± 6.29	22.96 ± 4.65	0.015	0.027	0.519
WHtR (cm)	0.48 ± 0.06	0.46 ± 0.07	0.47 ± 0.07	0.45 ± 0.06	<0.001	0.818	0.910
BMI (cm)	23.73 ± 4.98	22.16 ± 5.33	22.35 ± 4.98	20.22 ± 4.61	<0.001	0.170	0.893
pBMI (%)	73.24 ± 26.71	63.10 ± 29.01	66.18 ± 29.80	54.95 ± 31.15	<0.001	0.003	0.827
SBP (mmHg)	125.47 ± 10.12	109.18 ± 9.84	127.16 ± 12.69	106.93 ± 11.56	<0.001	0.048	0.600
DBP (mmHg)	80.75 ± 7.18	69.71 ± 5.96	82.06 ± 6.76	69.84 ± 6.25	<0.001	0.565	0.870
HR (bpm)	84.31 ± 14.97	80.39 ± 11.39	82.95 ± 13.27	79.16 ± 12.01	<0.001	0.017	0.438

Legend: Mean ± SD; SD: standard deviation; NT: normotensive; HT: hypertensive; WC: waist circumference; HC: hip circumference; MUAC: mid-upper arm circumference; WHtR: waist to height ratio; BMI: body mass index; pBMI: BMI percentile; NC: neck circumference; SBP: systolic blood pressure; DBP: diastolic blood pressure; HR: heart rate.

3.2. Relationship between Blood Pressure Measures and Obesity

Comparing blood pressure measures when participants were classified as obese using the standard pBMI cut-off at p95th (observed obesity) showed that the systolic blood pressure SBP and diastolic blood pressure DBP were higher ($p < 0.05$) in obese adolescents than non-obese adolescents. When individuals were classified as obese using pBMI cut-off at p85.2nd (predicted obesity), obese individuals had higher ($p < 0.001$) SBP and DBP than their non-obese counterparts. Observed obesity, defined by WHtR (0.481), showed higher SBP, DBP, and HR ($p < 0.05$) compared to their controls. Equally, obese individuals defined by HC, NC, MUAC, and NC showed higher SBP and DBP ($p < 0.05$) than their non-obese counterparts (Table 2).

Table 2. Comparison of blood pressure measures in children with or without obesity.

Observed Obesity by pBMI at 95th Percentile	**Non-Obese (967)**	**Observed Obesity (177)**	$t_{(1142)}$	**p-Value**
SBP (mmHg)	110.55 ± 12.36	116,29 ± 11.44	−5.749	<0.001
DBP (mmHg)	71.37 ± 7.61	73.16 ± 6.91	−2.912	0.004
HR (bpm)	80.67 ± 12.45	80.72 ± 10.78	−0.052	0.958
Predicted Obesity by pBMI at 85.2nd Percentile	**Non-Obese (781)**	**Predicted Obesity (352)**	$t_{(1133)}$	**p-Value**
SBP (mmHg)	109.7021	115.43 ± 11.99	7.350	<0.001
DBP (mmHg)	70.96 ± 7.63	73.18 ± 7.12	4.599	<0.001
HR (bpm)	80.67 ± 12.37	80.77 ± 11.87	0.131	0.896
Predicted Obesity by WHtR at 0.481	**Non-Obese (782)**	**Predicted Obesity (362)**	$t_{(1144)}$	**p-Value**
SBP (mmHg)	109.79 ± 12.08	114.99 ± 12.34	6.727	<0.001
DBP (mmHg)	70.91 ± 7.40	73.26 ± 7.57	4.912	<0.001
HR (bpm)	80.13 ± 12.35	81.86 ± 11.81	2.235	0.026
Predicted Obesity by WHtR at 0.5	**Non-Obese (855)**	**Predicted Obesity (289)**	$t_{(1142)}$	**p-Value**
SBP (mmHg)	110.1856 ± 12.27	115.16 ± 12.03	−5.986	<0.001
DBP (mmHg)	71.06 ± 7.46	73.38 ± 7.48	−4.551	<0.001
HR (bpm)	80.42 ± 12.39	81.43 ± 11.61	−1.210	0.226
Predicted Obesity by WC at 75.1 cm	**Non-Obese (787)**	**Predicted Obesity (357)**	$t_{(1134)}$	**p-Value**
SBP (mmHg)	109.16 ± 12.03	116.48 ± 11.68	9.619	<0.001
DBP (mmHg)	70.77 ± 7.52	73.58 ± 7.21	5.916	<0.001
HR (bpm)	80.73 ± 12.44	80.55 ± 11.69	−0.230	0.818

Table 2. *Cont.*

Predicted Obesity by HC at 92.2 cm	Non-Obese (705)	Predicted Obesity (438)	$t_{(1133)}$	*p*-Value
SBP (mmHg)	108.32 ± 11.72	116.49 ± 11.79	11.434	<0.001
DBP (mmHg)	70.69 ± 7.58	73.20 ± 7.19	5.532	<0.001
HR (bpm)	80.82 ± 12.48	80.46 ± 11.77	−0.491	0.628
Predicted Obesity by MUAC at 26 cm	**Non-Obese (513)**	**Predicted Obesity (249)**	$t_{(762)}$	***p*-Value**
SBP (mmHg)	105.57 ± 10.61	112.08 ± 12.35	7.521	<0.001
DBP (mmHg)	71.07 ± 7.56	72.77 ± 7.56	2.907	0.004
HR (bpm)	81.57 ± 12.62	82.88 ± 12.34	1.359	0.178
Predicted Obesity by NC at 30.6 cm	**Non-Obese (501)**	**Predicted Obesity (263)**	$t_{(764)}$	***p*-Value**
SBP (mmHg)	105.31 ± 11.17	112.32 ± 11.05	8.269	<0.001
DBP (mmHg)	70.99 ± 7.68	72.85 ± 7.27	3.217	0.001
HR (bpm)	82.23 ± 12.56	81.57 ± 12.45	−0.688	0.492

Legend: Mean ± SD; SD: standard deviation; SBP: systolic blood pressure; DBP: diastolic blood pressure; HR: heart rate. WC: waist circumference; HC: hip circumference; MUAC: mid-upper arm circumference; WHtR: waist to height ratio; pBMI: body mass index percentile; NC: neck circumference.

3.3. Relationship between Hypertension and Obesity

There was an association between obesity and hypertension (Table 3). However, the association was higher for obese individuals defined by the predicted obesity cut-off (χ^2 = 20.727; $p < 0.001$) than those defined by the observed obesity cut-off (χ^2 = 8.227; $p = 0.004$). Similarly, the odds of developing hypertension were greater using the predicted obesity cut-off (OR: 2.070) compared to the predicted obesity cut-off value (OR: 1.748).

Table 3. Relationship between obesity and hypertension.

		Normotensive	Hypertensive	Total	Chi-Square	OR	*p*-Value
Observed	Non-obese	817 (85.9)	150 (77.7)	967	8.227	1.748	0.004
Obesity	Obese	134 (14.1)	43 (22.3)	177			
(pBMI at p95th)	Total	951	193	1144			
Predicted	Non-obese	685 (72.0)	107 (55.4)	792	20.727	2.070	<0.001
Obesity	Obese	266 (28.0)	86 (44.6)	352			
(pBMI at p85th)	Total	951	193	1144			

3.4. Performance of Blood Pressure Measures to Screen Obesity

The ROC analysis of blood pressure parameters using obesity cut-offs is summarised in Table 4. The ROC curve for SBP, DBP, HR, SBP%, and DBP% using pBMI85.2 as reference, illustrated in Figure 1A, showed a low AUC for BP measures ranging between 0.504 and 0.639 with SBP having the highest AUC. The sensitivity and specificity were highest for DBP% (0.657) and SBP% (0.624), respectively. The cut-off values to predict hypertension by SBP, DBP, HR, SBP%, and DBP% were 111.17; 71.17; 77.17; 65.5, and 70.25, respectively.

As shown in Figure 1B, the ROC curve for SBP, DBP, HR, SBP%, and DBP% using pBMI95 as reference showed a low AUC for BP measures which ranged between 0.504 and 0.642. SBP had the highest AUC, sensitivity (0.670), and specificity (0.586). The cut-off values to predict hypertension by SBP, DBP, HR, SBP%, and DBP% were 112.17; 71.83; 77.17; 63.85, and 71.80, respectively.

Using WHtR of 0.481 as the reference showed a low AUC for BP measures with SBP having the highest AUC of 0.621. SBP% had the highest sensitivity (0.701) and SBP had the best specificity (0.529). The predicted cut-off values for hypertension were 110.5; 70.50; 77.50; 50.5, and 69.9 for SBP, DBP, HR, SBP%, and DBP%, respectively (Figure 1C).

The ROC curve for SBP, DBP, HR, SBP%, and DBP% using a WHtR of 0.5 as the reference, illustrated in Figure 1D, showed a low AUC for BP measures with SBP having the highest AUC of 0.632. SBP% had the highest sensitivity and specificity of 0.699 and 0.615, respectively. The predicted cut-off values for hypertension were 111.8; 70.17; 77.17; 65.5, and 78.5 for SBP, DBP, HR, SBP%, and DBP%, respectively (Figure 1D).

Table 4. Optimal cut-off, sensitivity, specificity, SE, and area under the ROC curves for blood pressure measures in predicting hypertension using pBMI and WHtR.

	Cut-Off	Sensitivity	1-Specificity	AUC	SE	95%CI
Ref: pBMI 85.2						
SBP	111.17	0.651	0.439	0.639	0.018	0.605–0.674
DBP	71.17	0.577	0.455	0.588	0.018	0.553–0.623
HR	77.17	0.597	0.465	0.504	0.018	0.468–0.540
SBP%	65.5	0.589	0.376	0.635	0.018	0.601–0.670
DBP%	70.25	0.657	0.553	0.560	0.018	0.524–0.595
Ref: pBMI 95						
SBP	112.17	0.670	0.434	0.642	0.022	0.599–0.686
DBP	71.83	0.563	0.456	0.569	0.022	0.526–0.612
HR	77.17	0.608	0.569	0.504	0.022	0.460–0.547
SBP%	63.85	0.670	0.414	0.641	0.022	0.598–0.685
DBP%	71.80	0.614	0.551	0.543	0.022	0.499–0.587
Ref: WHtR 0.48						
SBP	110.5	0.650	0.471	0.615	0.018	0.583–0.657
DBP	70.50	0.624	0.495	0.579	0.018	0.545–0.619
HR	77.50	0.611	0.549	0.538	0.018	0.500–0.574
SBP%	50.50	0.708	0.515	0.621	0.017	0.596–0.669
DBP%	69.90	0.662	0.559	0.557	0.018	0.527–0.602
Ref: WHtR 0.5						
SBP	111.83	0.613	0.443	0.620	0.019	0.583–0.657
DBP	70.17	0.642	0.517	0.582	0.019	0.545–0.619
HR	77.17	0.583	0.533	0.537	0.019	0.500–0.574
SBP%	65.50	0.699	0.385	0.632	0.019	0.596–0.669
DBP%	78.50	0.500	0.404	0.564	0.019	0.527–0.602

Legend: WHtR: waist to height ratio; pBMI: body mass index percentile; SBP: systolic blood pressure; DBP: diastolic blood pressure; HR: heart rate; AUC: area under the curve; SE: standard error; CI: confidence interval.

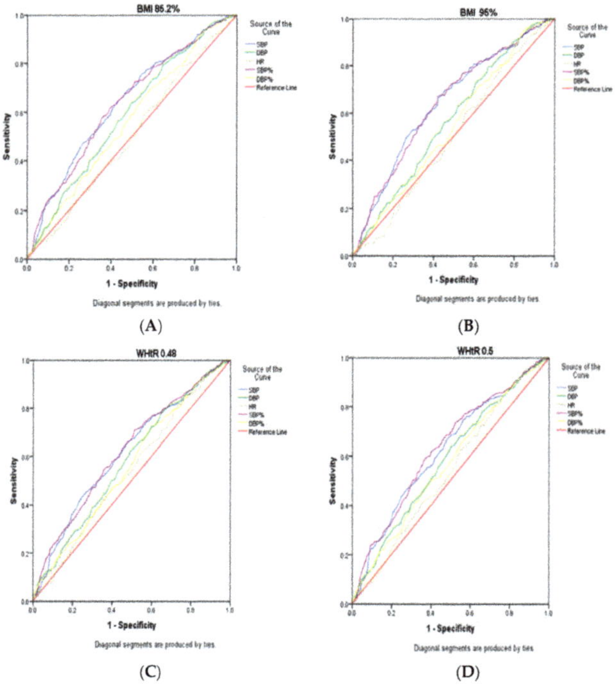

Figure 1. Receiver operating characteristic (ROC) curves for blood pressure measures using obesity measures (BMI% and WHtR) as reference (**A–D**).

Generally, obesity measures, both the WHO reference cut-off for WHtR of 0.5 and the CDC recommended pBMI% cut-off of 95%, as well as the new cut-offs for pBMI% of 85.2% and WHtR 0.48, performed poorly in predicting hypertension as the ROC curve showed the AUC below 70% for SBP percentile (SBP%) and DBP percentile (DBP%). The new cut offs for pBMI% and WHtR did not improve the sensitivity to predict hypertension as compared to the recommended cut-off values. The cut-off values for SBP% and DBP% from the ROC analysis ranged between the 60th and 70th percentile and were lower that the recommended value of 95 percentile.

4. Discussion

Various anthropometric indices used to screen obesity have been shown to predict hypertension in adulthood, among which a few have been suggested as an alternative for screening hypertension [16]. However, no consensus on anthropometric measures in screening childhood hypertension has been established. More so, with the increasing prevalence of hypertension in children and adolescents, there is a need for an alternative rapid diagnostic tool for hypertension in this population. This present study investigated the performance of obesity cut-offs established in our previous study to predict hypertension in adolescents in a South African population. The findings showed that the new obesity cut-off for pBMI and WHtR, as well as the CDC and WHO reference obesity cut-offs, had a poor performance in screening hypertension in this adolescent population, with a low AUC as well as low sensitivity and specificity. Hence, these obesity cut-offs had a poor discriminatory power for hypertension screening in adolescents.

Obesity has been considered a major risk factor for hypertension development both in adults [29] and children [30] and, therefore, may predict the presence of hypertension. However, the CDC and WHO recommended reference cut-off values of pBMI and WHtR, respectively, for obesity screening have been reported to differ across various race and ethnic population [31]. As such, this variation may affect the ability of these obesity screening tools to predict the diagnosis of hypertension across various ethnic populations. Based on the established obesity cut-off values for pBMI and WHtR in our previous report [21], which were lower than the recommend references by CDC and WHO, we hypothesized that they may predict hypertension differently. As such, we investigated the relationship of blood pressure and obesity defined by anthropometric cut-off values. Our findings showed that SBP and DBP were higher ($p < 0.05$) in obese adolescents compared to their non-obese counterparts. This is in accordance with previous studies which have shown obesity to be associated with hypertension in children [32,33] as well as in adolescents [34,35]. More so, obesity was strongly associated with hypertension ($p < 0.01$) with our predicted obesity (defined by pBMI85.2), showing a greater likelihood for hypertension (OR: 2.07; $p < 0.01$) than observed obesity defined by the recommended CDC reference of pBMI95 (OR: 1.748; $p < 0.01$). This finding suggests that the new pBMI% cut-off of 85.2% could be a better predictor of hypertension than the recommended CDC reference of 95%. The question which remains to be answered is if this new cut-off of obesity could serve as a bioindicator to predict hypertension screening in adolescents? This hypothesis was further investigated in this study using ROC analysis.

ROC analysis is commonly used to assess the performance of diagnostic screening tools as it provides the AUC as well as the sensitivity and specificity. In this case, the AUC evaluates the performance of anthropometric indicators for hypertension screening in adolescents. An AUC below 0.60 indicates a poor accuracy of obesity (pBMI and WHtR) cut-off in screening hypertension, whose occurrence may be random [36]. Moreover, an AUC above 0.9 is considered a good tool for its capacity to screen for hypertension. In this study, the AUC of pBMI and WHtR for screening DBP percentile and HR were <0.6 while the SBP percentile was 0.6 but ranged between 0.6 and 0.65. Similarly, the sensitivity of these cut-offs in predicting hypertension were between 50 and 60% for the HR and DBP percentile and between 60 and 70% for the SBP percentile. Equally, the sensitivity was low (<0.6) for all BP measures (DBP, SBP, and HR). There was no difference in the AUC, sensitivity, and specificity between the new cut-offs and the WHO and CDC reference cut-offs for pBMI and WHtR. Further, the cut-off values for blood pressure measures, as established from ROC analysis using anthropometric measures, were far below the recommended reference cut-off values for hypertension screening. This finding further confirmed obesity measures as a poor diagnostic tool for hypertension screening. This finding corroborates with a study involving Iranian children and adults which showed anthropometric indices with a poor accuracy for predicting elevated blood pressure and hypertension [37]. However, this contrasts with another study in Brazil which showed anthropometric indicators to have a good accuracy for high blood pressure screening in male adolescents [17]; meanwhile, studies in Spain [38] and Malaysia [16] showed anthropometric indices to have a moderate predictive capacity for high blood pressure in children and adolescents. Therefore, obesity measures may be considered a risk predictor rather than a screening tool for hypertension. This may be because obesity is an independent factor and not the sole risk factor for hypertension, as other factors such as high salts intake, alcohol, smoking, heredity, stress renal dysfunction, etc., present a risk for the development of hypertension [39,40]. This becomes even more complex as other factors, such as age, sex, and body composition, are related to blood pressure in children [41,42].

The strengths of this study, which supports these findings, are that a large population size was used and therefore backed by a strong predictive and statistical power. However, this study may be limited in that direct quantification of body fat was not used as a measure for obesity in predicting hypertension, considering that obesity is strongly associated with hypertension in the adult population. More so, other predictors of hypertension such as dyslipidaemia, smoking, hereditary, salt intake, etc., were not assessed. The study recruited predominately adolescents of Xhosa ethnicity and therefore generalization of the results to other populations in South Africa may not be applicable. The fact that the study was a cross-sectional study and other risk factors such as dietary intake, physical activity, and sedentary lifestyle, whose impact on hypertension depends on long-term monitoring, were not assessed limits the broader application of these findings.

5. Conclusions

Obesity established cut-offs for BMI percentile and WHtR in this populations showed poor performance in diagnosing hypertension even though they were strong predictors of hypertension. Therefore, obesity measures may not be sufficient to screen for hypertension in this adolescent population. However, obesity remains a significant risk to hypertension development and mandates regular monitoring in adolescents.

Author Contributions: Conceptualization, G.A.E., N.G. and B.N.N.-C.; Data curation, G.A.E., P.C.A. and B.N.N.-C.; Formal analysis, G.A.E.; Funding acquisition, B.N.N.-C.; Investigation, G.A.E. and P.C.A.; Validation, B.N.N.-C. and N.G.; Writing—original draft, G.A.E.; Writing—review and editing, G.A.E., P.C.A., N.G. and B.N.N.-C. All authors have read and agreed to the published version of the manuscript.

Funding: This study was funded by the South African National Research Foundation NRF-CPRR, Grant No: 106151 to Benedicta N. Nkeh-Chungag.

Institutional Review Board Statement: The study was approved by the Ethics Committee of University of Fort Hare (CH1011SCHU01) on 1 June 2016 and the Faculty of Health Sciences, Walter Sisulu University (Ref No: 112/2018) on 31 October 2018, South Africa.

Informed Consent Statement: Prior to enronment into the study, written informed consent was obtained from the parents or legal guardians of the children.

Data Availability Statement: All participant's data are kept confidential as per the South Africa National Data Protection Act. However, data may be made available upon reasonable request to the author of correspondence.

Acknowledgments: The authors are thankful to the management of the primary schools for granting permission for the study.

Conflicts of Interest: The authors declare no conflict of interest.

References

1. Mills, K.T.; Stefanescu, A.; He, J. The global epidemiology of hypertension. *Nat. Rev. Nephrol.* **2020**, *16*, 223–237. [CrossRef]
2. Urbina, E.M.; Khoury, P.R.; Bazzano, L.; Burns, T.L.; Daniels, S.; Dwyer, T.; Hu, T.; Jacobs, D.R., Jr.; Juonala, M.; Prineas, R.; et al. Relation of Blood Pressure in Childhood to Self-Reported Hypertension in Adulthood. The International Childhood Cardiovascular Cohort Consortium. *Hypertension* **2019**, *73*, 1224–1230. [CrossRef]
3. Azegami, T.; Uchida, K.; Tokumura, M.; Mori, M. Blood pressure tracking from childhood to adulthood. *Front. Pediatr.* **2021**, *9*, 785356. [CrossRef] [PubMed]
4. Noubiap, J.J.; Essouma, M.; Bigna, J.J.; Jingi, A.M.; Aminde, L.N.; Nansseu, J.R. Prevalence of elevated blood pressure in children and adolescents in Africa: A systematic review and meta-analysis. *Lancet Public Health* **2017**, *2*, 375–386. [CrossRef] [PubMed]
5. Song, P.; Zhang, Y.; Yu, J.; Zha, M.; Zhu, Y.; Rahimi, K.; Rudan, R. Global Prevalence of Hypertension in Children: A Systematic Review and Meta-analysis. *JAMA Pediatr.* **2019**, *173*, 1154–1163. [CrossRef]
6. Crouch, S.H.; Soepnel, L.M.; Kolkenbeck-Ruh, A.; Maposa, I.; Naidoo, S.; Davies, J.; Norris, S.A.; Ware, L.J. Paediatric hypertension in africa: A systematic review and meta-analysis. *EClinicalMedicine* **2022**, *43*, 101229. [CrossRef]
7. Kagura, J.; Ong, K.K.; Adair, L.S.; Pettifor, J.M.; Norris, S.A. Paediatric hypertension in South Africa: An underestimated problem calling for action. *SAMJ S. Afr. Med. J.* **2018**, *108*, 708–709. [CrossRef]
8. Matjuda, E.N.; Engwa, G.A.; Letswalo, P.B.; Mungamba, M.M.; Sewani-Rusike, C.R.; Nkeh-Chungag, B.N. Association of Hypertension and Obesity with Risk Factors of Cardiovascular Diseases in Children Aged 6–9 Years Old in the Eastern Cape Province of South Africa. *Children* **2020**, *7*, 25. [CrossRef]
9. Letswalo, B.P.; Schmid-Zalaudek, K.; Brix, B.; Matjuda, E.N.; Klosz, F.; Obernhumer, N.; Gaisl, M.; Engwa, G.A.; Sewani-Rusike, C.; Fredriksen, P.M.; et al. Cardiometabolic risk factors and early indicators of vascular dysfunction: A cross-sectional cohort study in South African adolescents. *BMJ Open* **2021**, *11*, e042955. [CrossRef]
10. Leggio, M.; Lombardi, M.; Caldarone, E.; Severi, P.; D'Emidio, S.; Armeni, M.; Bravi, V.; Maria, G.B.; Mazza, A. The relationship between obesity and hypertension: An updated comprehensive overview on vicious twins. *Hypertens. Res.* **2017**, *40*, 947–963. [CrossRef]
11. Yan, W.; Liu, F.; Li, X.; Wu, L.; Zhang, Y.; Cheng, Y.; Zhou, W.; Huang, G. Blood pressure percentiles by age and height for non-overweight Chinese children and adolescents: Analysis of the China Health and Nutrition Surveys 1991–2009. *BMC Pediatr.* **2013**, *13*, 195. [CrossRef]
12. Abolfotouh, M.A.; Sallam, S.A.; Mohammed, M.S.; Loutfy, A.A.; Hasab, A.A. Prevalence of elevated blood pressure and association with obesity in Egyptian school adolescents. *Int. J. Hypertens.* **2011**, *2011*, 952537. [CrossRef]
13. Pelegrini, A.; Silva, D.A.S.; de Lima Silva, J.M.F.; Grigollo, L.; Petroski, E.L. Anthropometric indicators of obesity in the prediction of high body fat in adolescents. *Rev. Paul. Pediatr.* **2015**, *33*, 56–62. [CrossRef]
14. Erdal, İ.; Yalçin, S.S.; Aksan, A.; Gençal, D.; Kanbur, N. How useful are anthropometric measurements as predictive markers for elevated blood pressure in adolescents in different gender? *J. Pediatr. Endocrinol. Metab.* **2020**, *33*, 1203–1211. [CrossRef]
15. Yoo, E.G. Waist-to-height ratio as a screening tool for obesity and cardiometabolic risk. *Korean J. Pediatr.* **2016**, *59*, 425–431. [CrossRef]
16. Cheah, W.L.; Chang, C.T.; Hazmi, H.; Kho, G.W.F. Using Anthropometric Indicator to Identify Hypertension in Adolescents: A Study in Sarawak, Malaysia. *Int. J. Hypertens.* **2018**, *2018*, 6736251. [CrossRef]
17. Borges, L.L.; Gerage, A.M.; Monteiro, L.Z.; Ulbrich, A.Z.; Silva, D. Proposition of Cutoff Points for Anthropometric Indicators to Identify High Blood Pressure in Adolescents. *Front. Nutr.* **2022**, *9*, 874047. [CrossRef]
18. Poh, B.K.; Ang, Y.N.; Yeo, G.S.; Lee, Y.Z.; Lee, S.T.; Chia, J.S.M.; Wee, S.B. Anthropometric indices, but not birth weight, are associated with high blood pressure risk among Malay adolescents in Kuala Lumpur. *Dialogues Health* **2022**, *1*, 100006. [CrossRef]
19. Prastowo, N.A.; Haryono, I.R. Elevated Blood Pressure and Its Relationship with Bodyweight and Anthropometric Measurements among 8–11-Year-Old Indonesian School Children. *J. Public Health Res.* **2020**, *9*, 1723. [CrossRef]
20. Hansen, M.L.; Gunn, P.W.; Kaelber, D.C. Underdiagnosis of hypertension in children and adolescents. *JAMA* **2007**, *298*, 874–879. [CrossRef]
21. Engwa, G.A.; Schmid-Zalaudek, K.; Anye, C.; Letswalo, P.B.; Anye, P.C.; Mungamba, M.M.; Sewani-Rusike, C.R.; Goswami, N.; Nkeh-Chungag, B.N. Assessment of Anthropometric Indices for Optimal Cut Offs for Obesity Screening in a South African Adolescent Population. *Biology* **2021**, *10*, 1118. [CrossRef] [PubMed]
22. De Onis, M.; Onyango, A.W.; Borghi, E.; Siyam, A.; Nishida, C.; Siekmann, J. Development of a WHO growth reference for school-aged children and adolescents. *Bull. World Health Organ.* **2007**, *85*, 660. [CrossRef] [PubMed]
23. Kuczmarski, R.J.; Ogden, C.L.; Guo, S.S. 2000 CDC growth charts for the United States: Methods and development. *Vital Health Stat.* **2002**, *11*, 1–190.
24. Ostchega, Y.; Prineas, R.J.; Paulose-Ram, R.; Grim, C.M.; Willard, G.; Collins, D. National health and nutrition examination survey 1999–2000: Effect of observer training and protocol standardization on reducing blood pressure measurement error. *J. Clin. Epidemiol.* **2003**, *56*, 768–774. [CrossRef]
25. Stewart, A.; Marfell-Jones, M.; Olds, T.; Ridder, H. *International Standards for Anthropometric Assessment*; ISAK: Lower Hutt, New Zealand, 2011.

26. Centers for Disease Control and Prevention. *A SAS Program for the 2000 CDC Growth Charts (Ages 0 to <20 Years)*; Division of Nutrition, Physical Activity, and Obesity, National Center for Chronic Disease Prevention and Health Promotion, Centers for Disease Control and Prevention: Atlanta, GA, USA, 2014.
27. Ashwell, M.; Hsieh, S.D. Six reasons why the waist-to-height ratio is a rapid and effective global indicator for health risks of obesity and how its use could simplify the international public health message on obesity. *Int. J. Food Sci. Nutr.* **2005**, *56*, 303–307. [CrossRef] [PubMed]
28. Kumar, R.; Indrayan, A. Receiver Operating Characteristic (ROC) Curve for Medical Researchers. *Indian Pediatr.* **2011**, *48*, 277–287. [CrossRef] [PubMed]
29. Jiang, S.Z.; Lu, W.; Zong, X.F.; Ruan, H.Y.; Liu, Y. Obesity and hypertension. *Exp. Ther. Med.* **2016**, *12*, 2395–2399. [CrossRef]
30. Brady, T.M. Obesity-Related Hypertension in Children. *Front. Pediatr.* **2017**, *25*, 197. [CrossRef]
31. Lear, S.A.; Toma, M.; Birmingham, C.L.; Frohlich, J.J. Modification of the relationship between simple anthropometric indices and risk factors by ethnic background. *Metabolism* **2003**, *52*, 1295–1301. [CrossRef]
32. Andrade, G.N.; Matoso, L.F.; Miranda, J.W.B.; Lima, T.F.; Gazzinelli, G.; Vieira, E.W. Anthropometric indicators associated with high blood pressure in children living in urban and rural areas. *Rev. Latino-Am. Enferm.* **2019**, *27*, e3150. [CrossRef]
33. Kaur, S.; Sachdev, H.; Dwivedi, S.N.; Lakshmi, R.; Kapil, U.; Sareen, N. Association of obesity with hypertension amongst school-age children belonging to lower income group and middle-income group in national capital territory of Delhi. *Indian J. Commun. Med.* **2013**, *38*, 175–179. [CrossRef]
34. Chori, E.; Hassidim, A.; Hartal, M.; Havakuk, O.; Flint, N.; Ziv-Baran, T.; Arbel, Y. Trends in Adolescents Obesity and the Association between BMI and Blood Pressure: A Cross-Sectional Study in 714,922 Healthy Teenagers. *Am. J. Hypertens.* **2015**, *28*, 1157–1163. [CrossRef] [PubMed]
35. Dulskiene, V.; Kuciene, R.; Medzioniene, J.; Benetis, R. Association between obesity and high blood pressure among Lithuanian adolescents: A cross-sectional study. *Ital. J. Pediatr.* **2014**, *40*, 102. [CrossRef] [PubMed]
36. Borges, L.S. Medidas de Acurácia diagnóstica na pesquisa cardiovascular. *Int. J. Cardiovasc. Sci.* **2016**, *29*, 218–222. [CrossRef]
37. Yazdi, M.; Assadi, F.; Qorbani, M.; Daniali, S.S.; Heshmat, R.E.; Motlagh, M.; Kelishadi, R. Validity of anthropometric indices in predicting high blood pressure risk factors in Iranian children and adolescents: CASPIAN-V study. *J. Clin. Hypertens.* **2020**, *22*, 1009–1017. [CrossRef]
38. Vaquero-Álvarez, M.; Molina-Luque, R.; Fonseca-Pozo, F.J.; Molina-Recio, G.; López-Miranda, J.; Romero-Saldaña, M. Diagnostic Precision of Anthropometric Variables for the Detection of Hypertension in Children and Adolescents. *Int. J. Environ. Res. Public Health* **2020**, *17*, 4415. [CrossRef]
39. Mamdouh, H.; Alnakhi, W.K.; Hussain, H.Y.; Ibrahim, G.M.; Hussein, A.; Mahmoud, I.; Alawadi, F.; Hassanein, M.; Abdullatif, M.; AlAbady, K.; et al. Prevalence and associated risk factors of hypertension and pre-hypertension among the adult population: Findings from the Dubai Household Survey, 2019. *BMC Cardiovasc. Disord.* **2022**, *22*, 18. [CrossRef]
40. Princewel, F.; Cumber, S.N.; Kimbi, J.A.; Nkfusai, C.N.; Keka, E.I.; Viyoff, V.Z.; Beteck, T.E.; Bede, F.; Tsoka-Gwegweni, J.M.; Akum, E.A. Prevalence and risk factors associated with hypertension among adults in a rural setting: The case of Ombe, Cameroon. *Pan Afr. Med. J.* **2019**, *34*, 147. [CrossRef]
41. Hu, J.; Ding, Z.; Han, D.; Hai, B.; Lv, H.; Yin, J.; Shen, H.; Gu, A.; Yang, H. Prevalence of hypertension and related risk factors among children and adolescents at three separate visits: A large school-based study in China. *Front. Pediatr.* **2022**, *10*, 976317. [CrossRef]
42. Bilal, M.; Haseeb, A.; Saeed, A.; Saeed, A.; Ghaffar, P. Prevalence and Risk Factors of Hypertension among children attending outpatient department of a tertiary care hospital in Karachi. *Cureus* **2020**, *12*, e7957. [CrossRef]

children

MDPI

Article

A Longitudinal Investigation on the Effects of Sodium and Potassium Intake on the Development of Hypertension and Abdominal Obesity from Childhood to Young Adulthood amongst Ellisras Rural Population, South Africa

Thato Tshepo Raphadu [1], Peter Modupi Mphekgwana [2], Moloko Matshipi [1] and Kotsedi Daniel Monyeki [1,*]

[1] Department of Physiology and Environmental Health, University of Limpopo, Sovenga 0727, South Africa; thatoraphadu@gmail.com (T.T.R.); moloko.matshipi@ul.ac.za (M.M.)

[2] Department of Research Administration and Development, University of Limpopo, Sovenga 0727, South Africa; peter.mphekgwana@ul.ac.za

* Correspondence: kotsedi.monyeki@ul.ac.za; Tel.: +27-152682953

Abstract: Background: Hypertension (HT) and obesity have both been on the rise in children. Each is associated with an increase in cardiovascular disease risk, and both track into adulthood. Objectives: This study aimed to identify the association of sodium intake (Na), potassium (K) intake, and sodium-to-potassium (Na/K) ratio with the development of HT and abdominal obesity amongst the Ellisras rural population over time. Methods: In this longitudinal study, data on dietary intake of Na and K were collected using a 24 h recall questionnaire from a total of 325 participants tracked from 1999 (5–12 years), 2001 (7–14 years), and 2015 (18–30 years). The averages of Na and K intake were analysed using local food tables and the South African Food Composition Database System (SAFOODS). In addition, blood pressure (BP) and anthropometric measurements (waist circumference (WC) and height) were also examined. Parametric (independent *t*-test) and Chi-square/Fishers exact tests were conducted to determine the difference between the years for numerical data and categorical variables. A generalised estimating equation (GEE) was used to assess the association of Na intake, K intake and their ratio on BP, WC, and waist-to-height ratio (WHtR). Results: Our results indicate a significant positive association between K intake and WHtR, and even though the model was adjusted for age and sex, there was still an association with WHtR. The Na/K ratio was associated with both BP measurements and abdominal obesity, respectively. Furthermore, Na/K was shown to be associated with an increased risk of developing HT and abdominal obesity. Conclusions: In our study, we observed that an increase in the Na/K ratio is a predictor of HT and abdominal obesity over time compared to Na and K alone. However, more studies are required to further prove this.

Keywords: hypertension; abdominal obesity; longitudinal study

1. Introduction

Hypertension and obesity have both been on the rise in children. Both track into adulthood and are linked to an increased risk of cardiovascular disease, which increases the prevalence of heart disease and its associated morbidity and mortality [1]. South Africa is an upper-middle-income country in which hypertension is a highly prevalent condition. In South Africa, hypertension was accountable for almost 47,000 fatalities in 2000. Since then, the prevalence has increased from 25% to more than 40% of the population [2]. According to a 2019 government study, around 41% of women and 11% of males aged 15 and above were obese [3]. Dietary intake has been identified as one of the most important risk factors for non-communicable diseases. High sodium and low potassium intakes are associated with the early development of chronic diseases (e.g., hypertension and obesity) [4].

Potassium is an essential mineral for human health [5,6]. It is essential for the normal functioning of cells and organs [7,8] due to its involvement in nerve transmission, muscle contractions, blood pressure regulation, and skeleton integrity [5,9–11]. Fruits, legumes, starchy roots and tubers, whole grains, and vegetables are the primary dietary sources of potassium [6,12,13]. The regulation of urinary potassium excretion is primarily responsible for the homeostatic control of serum potassium concentration. Reduced dietary intake of potassium [6], increased consumption of sodium [14,15] due to food processing, and high salt intake all have an effect on potassium utilization [16]. Dietary potassium intake is reduced due to potassium loss during food processing [17].

Evidence reveals a link between sodium consumption and blood pressure (BP) levels [18]. Excess sodium consumption has been associated with the development of hypertension and its cardiovascular consequences [19]. Reduced sodium consumption, on the other hand, not only lowers blood pressure and hypertension incidence but also lowers cardiovascular morbidity and death [20]. Observational studies, on the other hand, show that dietary potassium intake is inversely linked with blood pressure in certain studies but not in others [21]. It is known that sodium and potassium can affect blood pressure through several mechanisms. According to evidence, these micronutrients interact in a way that significantly contributes to the development of primary hypertension [22,23]. Modern Western diets—which are particularly heavy in sodium and low in potassium—have a biological interaction with the kidneys that causes the body to have excessive sodium levels and insufficient potassium concentrations. Vascular smooth muscle cell concentration is the first biological effect of these alterations, which are then followed by an increase in peripheral vascular resistance, increased pressure, and hypertension [22,23].

Obesity has lately attracted attention as another potential health-related result despite the distinct detrimental effects of increased sodium intake on hypertension [24]. Adult obesity and sodium intake are significantly correlated, according to several studies [25]. Furthermore, research from nationally representative populations in Australia, South Korea, and the UK revealed a substantial link between sodium and obesity in both adults and children [26]. The role of soft drink intake was investigated, despite the fact that the mechanisms underlying the link between sodium and obesity have not yet been thoroughly established [27]. According to this hypothesis, higher energy intake from sugar-sweetened beverages used to quench the thirst brought on by high sodium intake would contribute to the development of obesity [28]. In other words, salt intake is thought to be indirectly related to soft drink consumption, which in turn is related to obesity due to its impact on fluid intake [29]. Although there is sufficient evidence regarding Na and K, there are very few longitudinal studies in Africa studying the effect of sodium and potassium intake and their ratio on hypertension and abdominal obesity from childhood into young adulthood. Most of the studies were conducted in Europe, Asia, and North America. A study conducted in Africa by Saeid et al. 2018 only assessed the proportion frequency of Na and K [30] and did not track the changes in Na, K and NA/K ratio over time. The study conducted in Ellisras rural children by van Den Ende et al. (2014) and Mashiane et al. (2018) focused their study on the association between Body Mass Index (BMI) and dietary intake in children and young adults, respectively [31,32]. Studies regarding the association between micronutrients such as Na, K, and hypertension and abdominal obesity have received little attention, and the relationship between dietary intake and hypertension and abdominal obesity is inconsistent. This discrepancy highlights the need for such a study, especially in rural African populations, and the need to provide data from an African perspective and not only from Western and Eastern countries. Therefore, we designed this study to investigate the effect of sodium and potassium intake and their ratio on hypertension and abdominal obesity from childhood into young adulthood in Ellisras.

2. Materials and Methods

2.1. Sampling Procedure

This study is a component of the Ellisras Longitudinal Study (ELS), the regional specifics of which have been covered elsewhere [33]. At first, the ELS used a cluster sampling strategy. Basically, 22 schools were randomly selected from 68 schools in the

Ellisras rural region to participate in the study (10 preschool and 12 primary schools) [33]. Aged 3 to 10 years, baseline data were gathered in 1996. Subsequent examinations of dietary consumption, blood pressure, and anthropometric measures were performed in 1999, 2001, 2015, and 2018. Longitudinal data from 1999, 2001, and 2015 were analysed for this study. The ultimate number of participants was 325, of whom 164 women and 161 men had complete measurements (Figure 1).

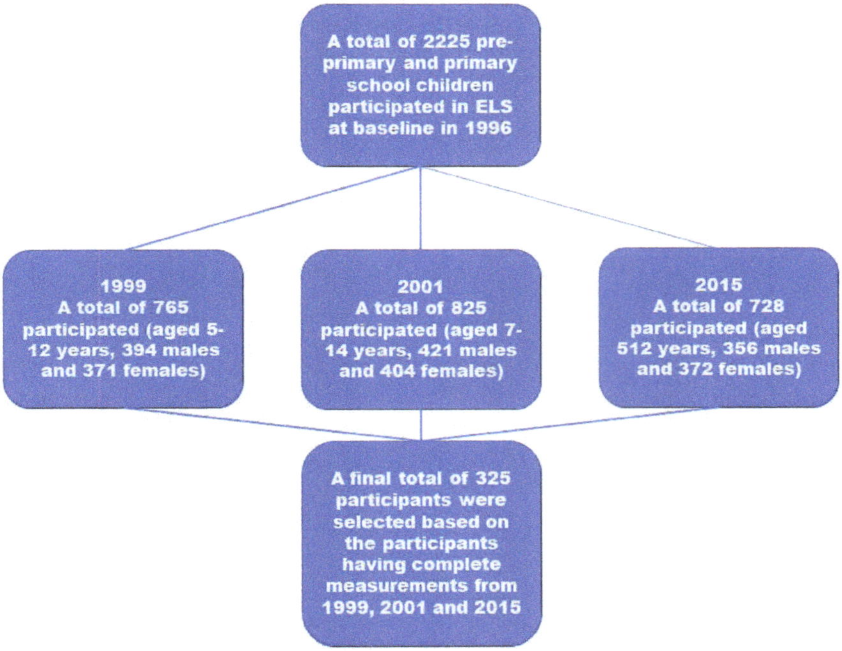

Figure 1. The flow chart of the study participants over the years.

2.2. Blood Pressure and Anthropometric Measurements

As described by Whelton et al. (2018) [34], blood pressure was monitored using Omron M2 blood pressure equipment. The National Heart, Lung, and Blood Institute (NHLBI) (2005) established cut-off points for high blood pressure or hypertension for age, sex, and height [35]. Boys' and girls' blood pressure was determined individually using the Merck manual medical calculator [36]. In contrast to adults, this method of computation calculates blood pressure for children and adolescents using percentiles. For adolescents over the age of eighteen and young adults, the cut-off points recommendations outlined by Whelton et al. (2018) were applied [34]. Weight, height, and waist circumference were measured anthropometrically in accordance with the International Society for the Advancement of Kinanthropometry's recommended practices [37]. The WC was divided by height in centimetres to obtain the waist-to-height ratio. Waist circumference (≥90th percentile for children and adolescents, ≥90 cm for males and ≥80 cm for females) and waist-to-height ratio (0.5) were used to characterise abdominal obesity [38,39].

2.3. Dietary Intake

A 24 h recall questionnaire was used to gather dietary data [40]. Participants were questioned by trained ELS field workers about their food consumption the previous day. All of the subjects' parents or caregivers were questioned about their nutritional consumption during the preceding 24 h. Using a pre-tested questionnaire and food models that represented typical serving sizes of regional meals, the estimated portion sizes of the items ingested were recorded in as much detail as feasible. Each participant's food consumption was recorded for an average of two days. Two food intakes—one for the week and the other for the weekend—were recorded [41]. This is due to the fact that what people eat during the week varies from what they eat on the weekends. Particularly on Saturdays, people tend to consume significantly more on the weekends than they do during the week [42]. The SAFOODS and local food tables were used to analyse the average sodium and potassium intakes [43], which were then compared to the recommended intakes for each nutrient as stated in the Consensus Study Report (2019) [44]. The WHO suggests using the sodium-to-potassium ratio of 1:1 (or 1) to determine the average molar sodium-to-potassium ratio [12]. The sodium and potassium intakes in this study were translated from milligrams to millimoles, since the suggested sodium/potassium ratio is expressed in moles. The molar sodium/potassium ratio was calculated using the calculation shown below [45]:

$$23 \text{ mg sodium} = 1 \text{ mmol sodium};$$

$$39 \text{ mg potassium} = 1 \text{ mmol potassium}$$

2.4. Statistical Analysis

In order to conduct the statistical analysis, the IBM SPSS Statistics software package (version 27.0) was employed. Descriptive statistics were produced for all the variables to provide frequencies (expressed as percentages), means, and standard deviations in order to describe and characterise the samples. The normality of the variables was evaluated by the Shapiro–Wilk test. To ascertain the variation in males and females throughout the years and between the years, parametric (one-way ANOVA), Chi-square, and Fisher's exact tests were used. To evaluate the relationship between sodium and potassium consumption, as well as their ratio with BP, WC, and WHtR, and further analyse the relationship with hypertension and abdominal obesity, generalised estimating equations (linear and binary logistics) were used. A p-value of ≤ 0.05 was chosen as the statistical significance for all tests.

3. Results

3.1. Characteristics of the Population

Table 1 indicates the characteristics of the population for each year stratified by sex, with a total of 325 participants. Although the dietary mean of Na consumption increased in both males and females with time, the rise was not significant in either males ($p > 0.05$) or females ($p > 0.05$). The dietary mean of K intake significantly decreased from the first year of assessment to the last year of assessment in both males ($p < 0.001$) and females ($p < 0.001$). Meanwhile, the dietary mean of the Na/K ratio significantly increased over the years in both males ($p < 0.001$) and females ($p < 0.001$). Furthermore, the mean of both SBP and DBP significantly increased from the first year of assessment to the last year of assessment, respectively, in both males ($p < 0.001$) and females ($p < 0.001$). The mean of abdominal indices (WC and WHtR) significantly increased over the years in both males ($p < 0.001$) and females ($p < 0.001$). Table 2 further shows that the dietary mean of the micro-nutrients (Na ($p < 0.05$), K intake ($p < 0.001$), and Na/K ratio ($p < 0.001$)) significantly increased over the years for the total population. In addition, the mean of BP measurements of the total population significantly increased from the first point of measurement, with SBP ($p < 0.001$) increasing to 120.75 mmHg and DBP ($p < 0.001$) increasing to 70.67 mmHg in the last year of measurement, respectively. The mean of abdominal indices significantly increased from the first measurement (54.44 cm-WC and 0.41-WHtR), with WC increasing to 79.0 cm and WHtR increasing to 0.48 in the last year.

Table 1. Characteristics of the population over the years, stratified by sex.

Year	1999		2001		2015	
	Males (n = 161)	**Females (n = 164)**	**Males (n = 161)**	**Females (n = 164)**	**Males (n = 161)**	**Females (n = 164)**
Variables	Mean ± SD	Mean ± SD	Mean ± SD	Mean ± SD	Mean ± SD	Mean ± SD
Age	8.78 ± 1.86	9.59 ± 1.75	10.78 ± 1.86	11.54 ± 1.75	24.79 ± 1.84	25.52 ± 3.13
Na (mg/d)	482.06 ± 593.49	532.84 ± 593.03	482.06 ± 593.49	535.82 ± 592.44	661.45 ± 988.80	7[illegible]7.18 ± 1082.93
K (mg/d)	1043.97 ± 836.65	1165.77 ± 963.26	1043.97 ± 836.65	1175.37 ± 968.15	756.28 ± 657.31 **	753.27 ± 678.56 **
Na/K ratio (mmol/d)	0.90 ± 1.03	0.89 ± 0.93	0.90 ± 1.03	0.89 ± 0.92	2.03 ± 2.75 **	1.92 ± 2.67 **
SBP (mmHg)	99.86 ± 11.71	102.17 ± 11.43	95.51 ± 9.90	99.39 ± 11.21	126.72 ± 13.11 **	1[illegible]5.00 ± 10.63 **
DBP (mmHg)	61.90 ± 9.89	62.92 ± 9.58	63.58 ± 7.21	64.66 ± 8.04	71.45 ± 10.17 **	69.93 ± 8.99 **
WC (cm)	53.93 ± 3.77	54.96 ± 4.69	56.65 ± 4.14	58.23 ± 4.30	74.53 ± 9.12 **	84.01 ± 15.39 **
WHtR	0.41 ± 0.02	0.40 ± 0.02	0.41 ± 0.04	0.41 ± 0.04	0.43 ± 0.08 **	0.52 ± 0.12 **
Height	130.36 ± 120.08	135.10 ± 10.68	138.51 ± 11.68	143.47 ± 10.77	173.54 ± 12.83 **	162.79 ± 10.29 **

n—number of individuals. Na—sodium, K—potassium, Na/K ratio—sodium/potassium ratio, SBP—systolic blood pressure, DBP—diastolic blood pressure, WHtR—waist-to-height ratio, mg/d—milligram per day, mmol/d—millimole per day, cm—centimetre, ** p-value < 0.001.

Table 2. Comparison of means of sodium, potassium intake, sodium-to-potassium ratio, systolic and diastolic blood pressure over the years.

Year	1999	2001	2015	p-Value
Variables	Mean ± SD	Mean ± SD	Mean ± SD	
Na (mg/d)	507.69 ± 592.89	509.27 ± 592.66	689.84 ± 1036.56	0.008 *
K (mg/d)	1105.43 ± 903.43	1110.48 ± 906.59	754.75 ± 667.16	0.000 **
Na/K ratio (mmol/d)	0.89 ± 0.98	0.89 ± 0.98	1.97 ± 2.70	0.000 **
SBP (mmHg)	101.02 ± 11.61	97.47 ± 10.74	120.75 ± 13.26	0.000 **
DBP (mmHg)	62.41 ± 9.73	64.12 ± 7.64	70.67 ± 9.60	0.000 **
WC (cm)	54.44 ± 4.27	57.47 ± 4.30	79.40 ± 13.53	0.000 **
WHtR	0.41 ± 0.24	0.41 ± 0.04	0.48 ± 0.11	0.000 **

Na—sodium, K—potassium, Na/K ratio—sodium/potassium ratio, SBP—systolic blood pressure, DBP—diastolic blood pressure, WHtR—waist-to-height ratio, SD—standard deviation, mg/d—milligram per day, mmol/d—millimole per day, cm—centimetre, * p-value < 0.05; ** p-value < 0.001.

Table 3 presents the most frequent food items in the diets of the Ellisras participants from 2001 and 2015. In 2001, maize or sorghum porridge was the most stable food among the Ellisras children, with jam being the least used item. In 2015, fried chicken with skin was the most used item, and pilchards were the least used food item among the Ellisras young adults.

Table 3. The most frequently used food items for the years 2001 and 2015 in Ellisras, from the most used to the least.

2001	2015 (%)
Maize porridge or Sorghum	Fried chicken with skin (23.8)
Sugar (white)	Pap (22.6)
Homemade bread	Cold drink (16.9)
Chicken	White sugar (14)
Spinach	Vetkoek (5.8)
Non-dietary creamer	Fried beef (4.7)
Beef	Peanut butter (4.4)
Red meat (from goats and wild animals)	Samp (2.6)
Tomato and onion	Yoghurt (2.4)
Cooked dry beans	Spinach (2.0)
Cold drink (mostly Coke)	Pilchards (0.5)
Peanut butter	
Sweets	
Mashontja (Mopani worms)	
Bananas and oranges	
Cow milk	
Jam	

Source adapted [18,19].

3.2. The Prevalence of Sodium, Potassium, Hypertension and Abdominal Obesity

Table 4 indicates a significant increase in the prevalence of hypertension from the first year to the third year in both males and females, respectively ($p < 0.001$). The prevalence of abdominal obesity according to WC significantly decreased in males, whereas in females, it increased from the first point of measurement to the third year ($p < 0.001$). In addition to this, both males' and females' prevalence of abdominal obesity according to WHtR significantly increased over the measured years ($p < 0.001$). The results in Table 5 show the prevalence of hypertension was 4.9% at the first measurement, and it significantly increased to 23.4% in the last year of assessment ($p < 0.001$). In addition, the prevalence of abdominal obesity according to WC significantly decreased by 5% from the first point of measurement to the last year of assessment ($p < 0.001$). Meanwhile, the prevalence of abdominal obesity significantly increased from 0.3% at the first measurement to 35.7% in the last year of measurement ($p < 0.001$).

Table 4. The prevalence of sodium, potassium, hypertension and abdominal obesity stratified by sex.

Year	1999		2001		2015	
Dependent variables	Males *n* (%)	Females *n* (%)	Males *n* (%)	Females *n* (%)	Males *n* (%)	Females *n* (%)
Na (mg/d) above the adequate intake	14 (8.7)	18 (11.0)	20 (12.4)	15 (9.1)	19 (11.8)	17 (10.4)
K (mg/d) below the adequate intake	156 (96.9)	153 (93.3)	153 (95.0)	150 (91.5)	155 (96.3)	163 (99.4) *
Na/K (mmol/d) above the recommended intake	57 (35.4)	56 (34.1)	58 (36.0)	56 (34.1)	84 (52.2) **	82 (50.0) *
Hypertension	8 (5.0)	8 (4.9)	4 (2.5)	6 (3.7)	36 (22.5) **	40 (24.2) **
Abdominal obesity according to WC	58 (36.0)	71 (43.6)	108 (67.1)	131 (79.9)	17 (10.6) **	96 (58.2) **
Abdominal obesity according to WHtR	0 (0)	1 (0.6)	4 (2.5)	10 (6.1)	19 (11.9) **	97 (58.8) **

n—number of individuals. Na—sodium, K—potassium, WC—waist circumference, WHtR—waist-to-height ratio, mg/d—milligram per day, mmol/d- millimole per day, * *p*-value < 0.05, ** *p*-value < 0.001.

Table 5. The prevalence of sodium, potassium, hypertension and abdominal obesity over the years.

Year	1999	2001	2015	Chi-Square *p*-Value
Dependant variables	*n* (%)	*n* (%)	*n* (%)	
Na (mg/d) above the adequate intake	32 (9.8)	35 (10.8)	36 (11.1)	0.868
K (mg/d) below the adequate intake	309 (95.1)	303 (93.2)	318 (97.8)	0.019 *
Na/K (mmol/d) above the recommended intake	113 (34.8)	114 (35.1)	166 (51.1)	<0.001 *
Hypertension	16 (4.9)	10 (3.1)	76 (23.4)	<0.001 **
Abdominal obesity according to WC	129 (39.8)	239 (73.5)	113 (34.8)	<0.001 **
Abdominal obesity according to WHtR	1 (0.3)	14 (4.3)	116 (35.7)	<0.001 **

n—number of individuals. Na—sodium, K—potassium, WC—waist circumference, WHtR—waist-to-height ratio, mg/d- milligram per day, mmol/d- millimole per day, * *p*-value < 0.05, ** *p*-value < 0.001.

3.3. The Association of Sodium Intake, Potassium Intake and Sodium-to-Potassium Ratio with Dependent Variables

The results in Table 6 show a significant positive association between potassium intake and WHtR ($\beta = 0.019$, (95% CL: 0.004, 0.034) p-value = 0.012), and even after age and gender were considered in the model, there was still an association with WHtR ($\beta = 0.018$, (95% CL: 0.002, 0.034) p-value = 0.024). The sodium-to-potassium ratio was significantly associated with SBP ($\beta = 4.326$, (95% CL: 2.056, 6.595) p-value < 0.001), DBP ($\beta = 2.028$, (95% CL: 0.703, 3.353) p-value = 0.003), WC ($\beta = 4.191$, (95% CL: 2.080, 6.302) p-value < 0.001) and WHtR ($\beta = 0.014$, (95% CL: 0.003, 0.026) p-value = 0.015), respectively.

However, when the model was adjusted for age and sex, an association was not found between the sodium-to-potassium ratio and blood pressure measurements and abdominal indices, respectively. Table 5 shows that in this study, it was further observed that the Na/K ratio increased the risk of developing hypertension (Exp β = 1.603, (95% CL: 1.164, 2.207) *p*-value = 0.004) and abdominal obesity (Exp β = 1.797, (95% CL: 1.207, 2.677) *p*-value = 0.004) compared to Na and K intake alone. However, when age and sex were taken into account, the sodium-to-potassium ratio was not found to increase the risk of developing hypertension and abdominal obesity, as shown in Table 7.

Table 6. The association of sodium intake, potassium intake and sodium-to-potassium ratio with blood pressure and abdominal indices using GEE (linear).

	Na (mg/d)			K (mg/d)			Na/K (mmol/d)		
Unadjusted	**β**	**95% CI**	***p*-value**	**β**	**95% CI**	***p*-value**	**β**	**95% CI**	***p*-value**
SBP	−0.507	−4.379, 3.366	0.798	2.033	−2.022, 6.088	0.326	4.326	2.056, 6.595	<0.001 **
DBP	−0.240	−2.287, 1.807	0.818	0.736	−1.738, 3.211	0.560	2.028	0.703, 3.353	0.003 *
WC	−0.735	−4.071, 2.600	0.666	2.314	−1.017, 5.644	0.173	4.191	2.080; 6.302	<0.001 **
WHtR	−0.001	−0.016, 0.013	0.866	0.019	0.004, 0.034	0.012 *	0.014	0.003, 0.026	0.015 *
Adjusted for age and sex									
SBP	0.915	−1.769, 3.599	0.504	0.579	−2.662, 3.821	0.726	0.967	−0.655, 2.589	0.243
DBP	0.311	−1.571, 2.194	0.746	0.283	−2.142, 2.709	0.819	0.798	−0.418, 2.014	0.198
WC	1.052	−0.668, 2.773	0.231	1.247	−0.642, 3.136	0.196	0.469	−0.701, 0.617	0.432
WHtR	0.004	−0.009, 0.017	0.57	0.018	0.002, 0.034	0.024 *	0.005	−0.005, 0.015	0.329

Na—sodium, K—potassium, Na/K ratio—sodium/potassium ratio, SBP—systolic blood pressure, DBP—diastolic blood pressure, WC—waist circumference, WHtR—waist-to-height ratio, mg/d—milligram per day, mmol/d—millimole per day, β—beta coefficient, * *p*-value < 0.05, ** *p*-value < 0.001.

Table 7. The risk measure of sodium, potassium intake and sodium-to-potassium ratio causing the development of hypertension and abdominal obesity according to WC and WHtR using GEE (binary logistic).

	Na (mg/d)			K (mg/d)			Na/K (mmol/d)		
Unadjusted	**Exp (β)**	**95% CI**	***p*-value**	**Exp (β)**	**95% CI**	***p*-value**	**Exp (β)**	**95% CI**	***p*-value**
Hypertension	0.959	0.590, 1.558	0.865	1.941	0.762, 4.941	0.164	1.603	1.164, 2.207	0.004 *
Abdominal obesity according to WC	1.131	0.730, 1.754	0.581	0.634	0.321, 1.254	0.190	0.912	0.684, 1.216	0.529
Abdominal obesity according to WHtR	1.066	0.579, 1.964	0.837	1.964	0.589, 6.545	0.272	1.797	1.207, 2.677	0.004 *
Adjusted for age and sex									
Hypertension	0.873	0.530, 1.437	0.593	0.679	0.283, 1.632	0.387	0.968	0.668, 1.403	0.864
Abdominal obesity according to WC	1.136	0.715, 1.805	0.588	0.698	0.362, 1.346	0.283	0.989	0.731, 1.338	0.943
Abdominal obesity according to WHtR	0.654	0.312, 1.372	0.262	0.702	0.221, 2.233	0.549	0.781	0.493, 1.239	0.295

Na—sodium, K—potassium, Na/K ratio—sodium/potassium ratio, WC—waist circumference, WHtR—waist-to-height ratio, mg/d—milligram per day, mmol/d—millimole per day, Exp β—exponential beta, * *p*-value < 0.05.

4. Discussion

The purpose of this study was to investigate the association of sodium and potassium intake and their ratio on hypertension and abdominal obesity in the same participants over time in Ellisras. The results of our study indicated that K was significantly associated with WHtR. The outcome of these results could be due to a lack of intake of fruits and vegetables in 2015 compared to 2001 (e.g., the sample population consumed bananas and oranges, which are high in potassium), resulting in a decline in the mean intake of K over the years. Most studies in this field that have been conducted over the past five years indicate that the K effect on obesity is a new topic, and more robust studies with better design are warranted [46], especially on abdominal obesity, as previous studies mostly focused on BMI and not abdominal indices such as WC and WHtR. It is unknown exactly how potassium consumption affects obesity/MetS. Central obesity is a component of metabolic syndrome, and the mechanisms of obesity and MetS are homogeneous. Potassium channel function and obesity are associated [47,48]. Our results further showed a significant association between Na/K ratio and both BP measurements (increasing SBP by 4.236 mmHg and DBP by 2.028 mmHg) and abdominal indices (increasing WC by 4.191 cm and WHtR by

0.0014 cm); the association was positive. Despite the lack of studies that could be compared to this study, Pereira et al. (2019) found independent associations of Na and K intake with BP, but when evaluated in a combined manner, as in the case of the Na/K ratio, the effect was potentiated [49].

Our results, on the one hand, showed no relationship between Na and BP measurements and abdominal indices. Although Na mean intake significantly increased over time, the increase was not significant enough to be associated with BP measurement and abdominal indices in this sample population of Ellisras.

Our study results revealed that the Na/K ratio is the predictor of HT and abdominal obesity compared to Na and K alone. This means that an increase in the Na/K ratio increases the risk of developing hypertension by 1.603-fold and abdominal pain by 1.797-fold over time. Due to the lack of sufficient studies, we could not compare this study with relevant studies. However, Ge et al. (2016) also discovered that the urine Na/K ratio was independently related to obesity and that a high Na/K ratio could increase the risk of obesity [50]. However, no link was discovered between the Na/K ratio (as determined by self-reporting) and obesity [51]. This contradicts the findings of this study, as the Na/K ratio was calculated using self-reported dietary data in both groups. Because the topic is new, data on the relationship between Na/K and abdominal obesity are inconsistent. On the other hand, our study is consistent with the literature, which states that the Na/K ratio outperforms Na and K, as individual predictors of BP change in various investigations. Most of the studies conducted in hypertensive patients [23] indicated that a higher Na/K ratio may lead to higher BP during follow-up [52]. Elevated levels of Na intake and inadequate levels of K intake may affect the development of hypertension [12]. To further agree with this literature, our results indicated an increase in Na over the years and a decrease in K over the years. In all the years, there were significant mean differences in males and females for the intake of potassium and the sodium-to-potassium ratio. In addition, there were significant mean differences in abdominal measurements for both males and females over the years.

On the other hand, BP measurements showed a significant mean difference in males and females over the years. Significant differences in the prevalence of hypertension, WC, and WHtR were observed in both genders over the years. Over time, there were significant variations in potassium, the sodium-to-potassium ratio, hypertension, WC, and WHtR prevalence. It must be noted that many of the participants were classified as underweight, as stated by van Den Ende et al. (2014) [31]. Therefore, the high prevalence of abdominal obesity according to WC compared to the prevalence of abdominal obesity according to WHtR 2001 might be due to bloating from malnourishment rather than visceral fat [41]. There was no significant change in the prevalence of sodium over time.

The change in political, social, and economic factors in South Africa has resulted in increased urbanisation and progress [31]. The increased accessibility, availability, and affordability of processed foods in South Africa are of concern, as these types of foods are generally considered to be high in fat, sugar or salt (sodium) [53,54]. Although our study found a lower intake of sodium and potassium, this is probably due to the fact that it was conducted in a rural settlement, because excessive intake of sodium and a deficient potassium intake may result in health issues.

There are a few limitations regarding this study. The first is the use of the 24 h recall questionnaire compared to the 24 h urinary excretion, as 24 h excretion is considered the golden standard method of obtaining data on sodium and potassium intake in population surveys and is more accurate than the 24 h recall questionnaire [30]. The second limitation is the small sample size and the lack of different ethnicities from different geographical regions. Future sodium excretion data over time, from childhood to adulthood within these regional areas, will be ideal for evaluating post-legislation salt intakes and their impact on public health [55]. Third, it does not include the socio-economic status of the sample population. In addition, the gap between the years is a limitation of this study. This was due to financial constraints. The strength of this study is the use of longitudinal data. The current study is valuable and informative regarding the status of sodium and potassium intake in a sample of South Africans. Tracking the dietary habits of children into adulthood is vital, as children with extremely high levels of sodium intake tend to maintain those levels for some time [56]. These could lead to the development of hypertension and abdominal obesity. Thus, the close monitoring of children is needed for better management of their health.

5. Conclusions

In conclusion, this current study showed a significant positive association between K and WHtR. In addition, there is a significant positive relationship between the Na/K ratio, BP measurements (SBP and DBP), and abdominal indices (WC and WHtR). An increase in the Na/K ratio was further found to increase the risk of developing HT and abdominal obesity over time. However, more data are crucial in establishing the effectiveness of Na reduction and the increase in K intake, and with the use of more valuable methods, we can better understand their effect on blood pressure and abdominal weight, especially in Africa. Longitudinal studies in children provide a tremendous global resource to direct prevention strategies for HT and abdominal obesity.

Author Contributions: Conceptualisation—K.D.M. and T.T.R.; Methodology—T.T.R.; Validation and Principal investigator—K.D.M.; Investigation—T.T.R.; Resources—K.D.M.; Collection of data—K.D.M., M.M. and P.M.M.; Writing—Original Draft Preparation—T.T.R.; Writing—Review and Editing—T.T.R., M.M. and P.M.M.; Supervision—K.D.M., M.M. and P.M.M. All authors have read and agreed to the published version of the manuscript.

Funding: This research received no external funding.

Institutional Review Board Statement: The ethical approval was granted by Turfloop research committee (protocol code TREC/88/2021: PG and 8 June 2021).

Informed Consent Statement: Informed consent was obtained from all subjects involved in the study.

Data Availability Statement: Data will be made available upon request.

Acknowledgments: We would like to acknowledge with sincere gratitude the principals of the selected schools and the fieldworkers who assisted with the collection of the data for this study. Gratitude to the Department of Physiology and Environmental Health, University of Limpopo, South Africa, and the ELS team.

Conflicts of Interest: The authors declare no conflict of interest.

References

1. Brady, T.M. Obesity-Related Hypertension in Children. *Front. Pediatr.* **2017**, *5*, 197. [CrossRef] [PubMed]
2. Ware, L.J.; Chidumwa, G.; Charlton, K.; Schutte, A.E.; Kowal, P. Predictors of hypertension awareness, treatment and control in South Africa: Results from the WHO-SAGE population survey (wave 2). *J. Hum. Hypertens.* **2019**, *33*, 157–166. [CrossRef] [PubMed]
3. National Department of Health; Statistics South Africa; South African Medical Research Council; ICF. *South Africa Demographic and Health Survey 2016*; Statistics South Africa (Stats SA) SAMRCSaI: Pretoria, South African, 2019.
4. Golpour-Hamedani, S.; Rafe, N.; Pourmasoumi, M.; Morteza Safavi, S.; Mohammadifard, N. Sodium and potassium intakes and adiposity among Iranian pre-adolescents and adolescents: A cross-sectional study. *Nutr. J.* **2022**, *21*, 23. [CrossRef] [PubMed]
5. Palmer, B.F. Regulation of potassium homeostasis. *Clin. J. Am. Soc. Nephrol.* **2015**, *10*, 1050–1060. [CrossRef]
6. Turck, D.; Bresson, J.L.; Burlingame, B.; Dean, T.; Fairweather-Tait, S.; Heinonen, M.; Hirsch-Ernst, K.I.; Mangelsdorf, I.; McArdle, H.; Neuhäuser-Berthold, M. Dietary reference values for potassium. *EFSA J.* **2016**, *14*, e04592.
7. Weaver, C.M. Potassium and health. *Adv. Nutr.* **2013**, *4*, 368S–377S. [CrossRef]
8. Viera, A.J.; Wouk, N. Potassium Disorders: Hypokalemia and Hyperkalemia. *Am. Fam. Physician* **2015**, *92*, 487–495.
9. Salomonsson, M.; Brasen, J.C.; Sorensen, C.M. Role of renal vascular potassium channels in physiology and pathophysiology. *Acta Physiol.* **2017**, *221*, 14–31. [CrossRef]
10. Kong, S.H.; Kim, J.H.; Hong, A.R.; Lee, J.H.; Kim, S.W.; Shin, C.S. Dietary potassium intake is beneficial to bone health in a low calcium intake population: The Korean National Health and Nutrition Examination Survey (KNHANES) (2008–2011) *Osteoporos. Int.* **2017**, *28*, 1577–1585. [CrossRef]

11. Newberry, S.J.; Chung, M.; Anderson, C.A.M.; Chen, C.; Fu, Z.; Tang, A.; Zhao, N.; Booth, M.; Marks, J.; Hollands, S.; et al. AHRQ comparative effectiveness reviews. In *Sodium and Potassium Intake: Effects on Chronic Disease Outcomes and Risks*; AHRQ Comparative Effectiveness Reviews; Agency for Healthcare Research and Quality (US): Rockville, MD, USA, 2018.
12. World Health Organisation. *Guideline: Potassium Intake for Adults and Children*; World Health Organisation: Geneva, Switzerland, 2012.
13. Broadley, M.R.; White, P.J. Eats roots and leaves. Can edible horticultural crops address dietary calcium, magnesium and potassium deficiencies? *Proc. Nutr. Soc.* **2010**, *69*, 601–612. [CrossRef]
14. Du, S.; Wang, H.; Zhang, B.; Popkin, B.M. Dietary potassium intake remains low and sodium Iitake remains high, and most sodium is derived from home food preparation for Chinese adults, 1991–2015 trends. *J. Nutr.* **2020**, *150*, 1230–1239. [CrossRef] [PubMed]
15. Rauber, F.; Da Costa Louzada, M.L.; Steele, E.M.; Millett, C.; Monteiro, C.A.; Levy, R.B. Ultra-processed food consumption and chronic non-communicable diseases-related dietary nutrient profile in the UK (2008–2014). *Nutrients* **2018**, *10*, 587. [CrossRef] [PubMed]
16. Brown, I.J.; Tzoulaki, I.; Candeias, V.; Elliott, P. Salt intakes around the world: Implications for public health. *Int. J. Epidemiol.* **2009**, *38*, 791–813. [CrossRef] [PubMed]
17. Sun, H.; Weaver, C.M. Rise in potassium deficiency in the US population links to agriculture practices and dietary potassium deficits. *J. Agric. Food Chem.* **2020**, *68*, 11121–11127. [CrossRef]
18. Mente, A.; O'Donnell, M.J.; Rangarajan, S.; McQueen, M.J.; Poirier, P.; Wielgosz, A.; Morrison, H.; Li, W.; Wang, X.; Di, C.; et al. Association of urinary sodium and potassium excretion with blood pressure. *N. Engl. J. Med.* **2014**, *371*, 601–611. [CrossRef]
19. Grillo, A.; Salvi, L.; Coruzzi, P.; Salvi, P.; Parati, G. Sodium intake and hypertension. *Nutrients* **2019**, *11*, 1970. [CrossRef]
20. Whelton, P.K.; He, J. Health effects of sodium and potassium in humans. *Curr. Opin. Lipidol.* **2014**, *25*, 75–79. [CrossRef]
21. Zhang, Z.; Cogswell, M.; Gillespie, C.; Fang, J.; Loustalot, F.; Dai, S.; Carriquiry, A.L.; Kuklina, E.V.; Hong, Y.; Merritt, R.; et al. Association between usual sodium and potassium intake and blood pressure and hypertension among U.S. adults: NHANES 2005–2010. *PLoS ONE* **2013**, *8*, e75289.
22. Adrogué, H.J.; Madias, N.E. Sodium and potassium in the pathogenesis of hypertension. *N. Engl. J. Med.* **2007**, *356*, 1966–1978. [CrossRef]
23. Perez, V.; Chang, E.T. Sodium-to-potassium ratio and blood pressure, hypertension, and related factors. *Adv. Nutr.* **2014**, *5*, 712–741. [CrossRef]
24. He, F.J.; Li, J.; Macgregor, G.A. Effect of longer-term modest salt reduction on blood pressure: Cochrane systematic review and meta-analysis of randomised trials. *BMJ* **2013**, *346*, f1325. [CrossRef] [PubMed]
25. Drenjancevic-Peric, I.; Jelakovic, B.; Lombard, J.H.; Kunert, M.P.; Kibel, A.; Gros, M. High-salt diet and hypertension: Focus on the renin-angiotensin system. *Kidney Blood Press. Res.* **2011**, *34*, 1–11. [CrossRef] [PubMed]
26. Millen, A.M.; Norton, G.R.; Majane, O.H.; Maseko, M.J.; Brooksbank, R.; Michel, F.S.; Snyman, T.; Sareli, P.; Woodiwiss, A.J. Insulin resistance and the relationship between urinary Na+/K+ and ambulatory blood pressure in a community of African ancestry. *Am. J. Hypertens.* **2013**, *26*, 708–716. [CrossRef]
27. Hedayati, S.S.; Minhajuddin, A.T.; Ijaz, A.; Moe, O.W.; Elsayed, E.F.; Reilly, R.F.; Huang, C.L. Association of urinary sodium/potassium ratio with blood pressure: Sex and racial differences. *Clin. J. Am. Soc. Nephrol.* **2012**, *7*, 315–322. [CrossRef]
28. Michel, F.S.; Norton, G.R.; Majane, O.H.; Badenhorst, M.; Vengethasamy, L.; Paiker, J.; Maseko, M.J.; Sareli, P.; Woodiwiss, A.J. Contribution of circulating angiotensinogen concentrations to variations in aldosterone and blood pressure in a group of African ancestry depends on salt intake. *Hypertension* **2012**, *59*, 62–69. [CrossRef] [PubMed]
29. Huggins, C.E.; O'Reilly, S.; Brinkman, M.; Hodge, A.; Giles, G.G.; English, D.R.; Nowson, C.A. Relationship of urinary sodium and sodium-to-potassium ratio to blood pressure in older adults in Australia. *Med. J. Aust.* **2011**, *195*, 128–132. [CrossRef]
30. Saeid, N.; Elmzibri, M.; Hamrani, A.; Latifa, Q.; Belghiti, H.; El Berri, H.; Benjeddou, K.; Bouziani, A.; Benkirane, H.; Taboz, Y.; et al. Assessment of Sodium and Potassium Intakes in Children Aged 6 to 18 Years by 24 h Urinary Excretion in City of Rabat. *Morocco. J. Nutr. Metab.* **2018**, *2018*, 8687192. [CrossRef]
31. Ende, C.V.D.; Twisk, J.; Monyeki, K. The relationship between BMI and dietary intake of primary school children from a rural area of South Africa: The Ellisras longitudinal study. *Am. J. Hum. Biol.* **2014**, *26*, 701–706. [CrossRef]
32. Mashiane, J.T.; Monyeki, K.D.; Kengne, A.P.; Nkwana, R.M.; Monyeki, M.S. Ellisras Longitudinal Study 2017: The relationship between dietary intake and body mass index among young rural adults in South Africa aged 18 to 30 years (ELS 18). *Cardiovasc. J. Afr.* **2018**, *29*, 301–304. [CrossRef]
33. Monyeki, K.D.; Van Lenthe, F.J.; Steyn, N.P. Obesity: Does it occur in African children in a rural community in South Africa? *Int. J. Epidemiol.* **1999**, *28*, 287–292. [CrossRef]
34. Whelton, P.K.; Carey, R.M.; Aronow, W.S. 2017ACC/AHA/AAPA/ABC/ACPM/AGS/APhA/ASH/ASPC/NMA/PCNA Guideline for the Prevention, Detection, Evaluation, and Management of High Blood Pressure in Adults. *J. Am. Coll. Cardiol.* **2018**, *7*, 68–74.
35. National Heart, Lung and Blood Institute. *The Fourth Report on the Diagnosis, Evaluation, and Treatment of High Blood Pressure in Children and Adolescents*; National Heart, Lung and Blood Institute: Bethesda, MD, USA, 2005.
36. Merck Manual-Medical Calculator. Blood Pressure Percentile. 2018. Available online: www.merckmanual.com/medical-calculators/BloodPressureBoys.htm (accessed on 15 May 2021).
37. Norton, K.; Olds, T. *Anthropometrics*; University of New South Wales Press: Sydney, Australia, 1996.

38. Vijayanchali, S.S. Prevalence of obesity in the rural adult population: An empirical assessment. *J. Res. Ext. Dev.* **2014**, *2*, 7–11.
39. Yoo, E. Waist-to-height ratio as a screening tool for obesity and cardiometabolic risk. *Korean J. Pediatr.* **2016**, *59*, 425–431. [CrossRef]
40. Langenhoven, M.L.; Conradie, P.J.; Wolmarans, P.; Faber, M. *MRC Food Quantities Manual*, 2nd ed.; Parow Research Institute for Nutritional Diseases, South African Medical Research Council: Cape Town, South Africa, 1991.
41. Raphadu, T.T.; Matshipi, M.; Mphekgwana, P.M.; Monyeki, K.D. Assessing the association of sodium, potassium intake and sodium/potassium ratio on blood pressure and central adiposity measurements amongst Ellisras undernourished, rural children aged 5–13 years: South Africa. *Children* **2022**, *9*, 422. [CrossRef] [PubMed]
42. Ann, R. Weekend-weekday differences in diet among U.S. adults, 2003–2012. *Ann. Epidemiol.* **2016**, *26*, 57–65. [CrossRef]
43. Wolmarans, P.; Kunneke, E.; Laubscher, R. Use of the South African Food Composition Database System (SAFOODS) and its products in assessing dietary intake data: Part II. *S. Afr. J. Clin. Nutr.* **2009**, *22*, 59–67. [CrossRef]
44. National Health Institute. *Consensus Study Report. Dietary Reference Intake for Sodium and Potassium*; National Health Institute: Ottawa, ON, Canada, 2019.
45. Department of Health and Ageing; National Health and Medical Research Council. *Nutrient Reference Values for Australia and New Zealand*; Department of Health and Ageing: Canberra, Australia, 2006.
46. Lee, H.; Lee, J.; Hwang, S.-S.; Kim, S.; Chin, H.J.; Han, J.S.; Heo, N.J. Potassium intake and the prevalence of metabolic syndrome: The Korean National Health and Nutrition Examination Survey 2008–2010. *PLoS ONE* **2013**, *8*, e55106. [CrossRef]
47. Climent, B.; Simonsen, U.; Rivera, L. Effects of obesity on vascular potassium channels. *Curr. Vasc. Pharmacol.* **2014**, *12*, 438–452. [CrossRef] [PubMed]
48. Crunkhorn, S. Metabolic disease: Potassium channel blocker prevents obesity. *Nat. Rev. Drug Discov.* **2013**, *12*, 577. [CrossRef] [PubMed]
49. Pereira, T.S.S.; Mill, J.G.; Griep, R.H.; Sichieri, R.; Molina, M.C.B. Effect of urinary sodium-to-potassium ratio change on blood pressure in participants of the longitudinal health of adult's study—ELSA-Brasil. *Medicine* **2019**, *98*, e16278. [CrossRef]
50. Ge, Z.; Zhang, J.; Chen, X.; Yan, L.; Guo, X.; Lu, Z.; Xu, A.; Ma, J. Are 24 h urinary sodium excretion and sodium: Potassium independently associated with obesity in Chinese adults? *Public Health Nutr.* **2016**, *19*, 1074–1080. [CrossRef] [PubMed]
51. Murakami, K.; Livingstone, M.B.E.; Sasaki, S.; Uenishi, K.; The Japan Dietetic Students' Study for Nutrition and Biomarkers Group. The ability of self-reported estimates of dietary sodium, potassium and protein to detect an association with general and abdominal obesity: Comparison with the estimates derived from 24 h urinary excretion. *Br. J. Nutr.* **2015**, *113*, 1308–1318. [CrossRef]
52. Okayama, A.; Okuda, N.; Miura, K.; Okamura, T.; Hayakawa, T.; Akasaka, H.; Ohnishi, H.; Saitoh, S.; Arai, Y.; Kiyohara, Y.; et al. Dietary sodium-to-potassium ratio as a risk factor for stroke, cardiovascular disease and all-cause mortality in Japan: The NIPPON DATA80 cohort study. *BMJ Open* **2016**, *6*, e011632. [CrossRef]
53. Feeley, A.B.; Norris, S.A. Added sugar and dietary sodium intake from purchased fast food, confectionery, sweetened beverages and snacks among Sowetan adolescents. *S. Afr. J. Child Health* **2014**, *8*, 88–91. [CrossRef]
54. Moodley, G.; Christofdes, N.; Norris, S.A.; Achia, T.; Hofman, K.J. Peer reviewed: Obesogenic environments: Access to and advertising of sugar-sweetened beverages in Soweto, South Africa, 2013. *Prev Chronic Dis.* **2015**, *12*, 140559. [CrossRef] [PubMed]
55. Swanepoela, B.; Schutte, A.E.; Cockeran, M.; Steyn, K.; Wentzel-Viljoen, E. Sodium and potassium intake in South Africa: An evaluation of 24-hour urine collections in a White, Black and Indian population. *J. Am. Soc. Hypertens.* **2016**, *10*, 829–837. [CrossRef]
56. Patterson, E.; Wärnberg, J.; Kearney, J.; Sjöström, M. The tracking of dietary intakes of children and adolescents in Sweden over six years: The European Youth Heart Study. *Int. J. Behav. Nutr. Phys. Act.* **2009**, *6*, 91. [CrossRef]

children

MDPI

Article

Child Nutrition Outcomes and Maternal Nutrition-Related Knowledge in Rural Localities of Mbombela, South Africa

Lucy Nomsa Masilela [1] and Perpetua Modjadji [1,2,*]

[1] Department of Public Health, School of Health Care Sciences, Sefako Makgatho Health Sciences University, 1 Molotlegi Street, Ga-Rankuwa, Pretoria 0208, South Africa

[2] Non-Communicable Disease Research Unit, South African Medical Research Council, Cape Town 7505, South Africa

* Correspondence: perpetua.modjadji@mrc.ac.za

Abstract: Poor nutrition outcomes among children have become one of the major public health concerns in South Africa, attributed to poor feeding practices and maternal nutrition-related knowledge with conflicting data. In view of this, a cross-sectional study was conducted to determine the association of nutrition outcomes of children aged under two years with feeding practices and maternal nutrition-related knowledge in Mbombela, South Africa. Mothers' nutrition-related knowledge was estimated using an adapted structured questionnaire on colostrum, continued breastfeeding, diarrhea prevention and treatment using oral rehydration solution, immunization, and family planning, and scored as excellent (80–100), good (60–79), average (40–59), and fair (0–39). This was along with questions on socio-demographic factors and obstetric history, as well as anthropometric measurements. Child nutrition outcomes were estimated by WHO classification using z-scores for stunting (length-for-age (LAZ)), underweight (weight-for-age (WAZ)), and thinness (body mass index-for-age (BAZ)). Using STATA 17, 400 pairs of children (8 ± 6 months) and their mothers (29 ± 6 years) participated in the study and were living in a poor socio-economic status environment. Half of children were stunted (50%) and over half (54%) were obese, while mothers were underweight (39%) and overweight (34%). In addition to one third of mothers reporting obstetric complications, two thirds, initiated breastfeeding within one hour of delivery, 30% exclusively breastfed, 48% introduced early complementary feeding, and 70% practiced mixed feeding. Twenty-eight percent (28%) of mothers had fair nutrition-related knowledge, while 66% had average knowledge, 6% good knowledge, and none of the mothers had excellent knowledge. A chi-square test showed that mothers' nutrition-related knowledge was significantly associated with child stunting. The final hierarchical logistic regression showed significant associations of stunting with mothers' nutrition-related knowledge (average: AOR = 1.92, 95%CI: 1.12–3.29), child's age (6–11 months: AOR = 2.63, 95%CI: 1.53–4.53 and 12–23 months: AOR = 3.19, 95%CI: 1.41–7.25), and education (completing Grade 12: AOR = 0.36, 95%CI: 0.15–0.86). Contextual and intensified interventions on continued education for mothers to gain accurate information on nutrition-related knowledge and feeding practices could ultimately enhance child nutrition outcomes in poorer settings. Efforts should therefore be made to ensure that nutrition knowledge is appropriately provided based on the phases of child growth from 0 to 2 years, even beyond infancy into school age.

Keywords: nutrition outcomes; nutrition knowledge; feeding practices; child–mother pairs; South Africa

1. Introduction

Child nutrition outcomes (i.e., stunting, underweight, and thinness) are key to tracking the nutrition and health status of children in a population [1]. In particular, stunting (i.e., linear growth retardation or low length-for-age (LAZ)) is affecting 21.9% of children under five years, on a global level, and 41% in Sub-Saharan Africa (SSA) [2]. Stunting is common in the first two years of life and is an indicator of chronic undernutrition [3]. In developing countries, the predominance of stunting is attributed to dietary quality, which ultimately interrupts early linear growth in stages of life into adulthood, affecting body size and intellectual functioning [4,5]. In the long term, stunting among children predisposes them to short adult stature, reduced lean body mass, less schooling, diminished intellectual functioning, and overweight or obesity later in life as well as a higher risk of coronary heart disease, stroke, hypertension, and type 2 diabetes and unproductivity later in life [4,6–9].

In South Africa, child nutrition outcomes remain a public health concern [10–13]. Approximately 27% to 32.3% of children under five years in the country are stunted, followed by underweight (9%) and wasted (2.5%) [11]. Stunting, being a multifactorial phenomenon, is attributable to factors such as poor socio-economic status, poor maternal health and nutrition, inadequate infant and young child feeding (IYCF), and micronutrient deficiencies [14]. Appropriate IYCF entails the initiation of breastfeeding within one hour of birth, exclusive breastfeeding (EBF) for six months, a continuation of breastfeeding for up to two years and beyond, as well as timely introduction of complementary foods [15,16]. According to the South African Demographic and Health Survey (SADHS), the percentage of exclusive breastfeeding was 32% in 2016, while 19% of children (12–23 months) were breastfed. Recent studies have estimated stunting among children aged under five years as between 45.3% to 55% [17,18] and associated delayed introduction of solid foods with stunting [11] due to inappropriate food supplementation [17]. Other studies conducted in Burkina Faso [19] and Ethiopia [20] reported that stunting is worsened by breastfeeding continuing with proper complementary feeding specific for age, making infants beyond six months vulnerable to poor nutrition outcomes [21].

Several developing countries have adopted a multisectoral approach and implemented specific and sensitive nutrition interventions to reduce childhood stunting [22]. Further, most countries are prioritizing policies and programs that improve mothers' ability to provide optimal care until the age of two years, a critical period for child growth and cognitive development and the origin of non-communicable diseases (NCDs) [22–24]. This is because the availability of parental resources in the first two years of a child's life influences proper growth and cognitive development [3–5]. For instance, a mother's knowledge in relation to food choices, feeding, health seeking, and her education has been studied [9–15]. According to the UNICEF Model of Care, the fundamental role of care in child nutrition indicates that caregivers require education, time, and support to provide adequate care [25]. Maternal knowledge and understanding of the aspects of basic nutrition and healthcare determine the care they give to their children [22,26,27]. It is worth noting that nutrition-related knowledge may be obtained from various sources [27], and if correct and put to practice, the nutrition-related knowledge may result in good child nutrition outcomes but, on the contrary, poor nutritional outcomes (i.e., stunting, underweight, and thinness) may result if incorrect information is put into practice [28]. Nonetheless, inconsistencies of the association of poor nutrition outcomes, nutrition-related knowledge, and feeding practices have been reported [29–33]. For instance, other studies reported significant associations [27,29,30], while others showed that adequate nutrition-related knowledge is not always translated into appropriate feeding practices [21,31,34,35].

The relationship of child nutrition outcomes and maternal nutrition-related knowledge in the South African context remains unclear. Few studies in the country have investigated several aspects of mothers regarding feeding knowledge [31], beliefs [36], practices [37], or dietary diversity [38] in relation to child nutrition outcomes. Nonetheless, stunting remains relatively high among South African children [17,18], and different forms of stunting and overweight/obesity in the same individuals [8,39], households, and settings coexist [38,40–43]. In view of this, the current study determined the association of nutrition outcomes of children aged under two years with feeding practices and maternal nutrition-related knowledge in Mbombela, South Africa. We hypothesize that in rural Mbombela, there is a significant relationship among feeding practices, nutrition-related knowledge, and child nutrition outcomes, especially with stunting. South Africa has promulgated several policies regarding feeding of children, including the Mother Baby Friendly Initiative (MBFI) and IYCF program [24,44].

Therefore, efforts to translate nutrition-related knowledge into appropriate action in an impoverished environment might be of help to mothers [45].

2. Materials and Methods

2.1. Study Design and the Conceptual Frameworks

This paper is an excerpt from a Master of Public Health dissertation written by the first author, which determined maternal feeding practices and nutritional knowledge, as well as the nutritional status of infants in community health facilities of the Mbombela sub-district in Mpumalanga Province. The entire study used a cross-sectional study design collecting data from mothers and their children under two years on anthropometry, obstetric history, socio-demographic status, feeding practices, and nutrition knowledge. The study was conducted from September 2021 to December 2021. From this umbrella study, the first published paper was "Evidence of Concurrent Stunting and Obesity among Children under two Years from Socio-Economically Disadvantaged Backgrounds in the Era of the Integrated Nutrition Programme in South Africa" [13]. The current paper is focused on child nutrition outcomes and maternal nutrition-related knowledge.

The methods used in this study have been described in detail in the first paper [13]. In a nutshell, the study was anchored on the Bronfenbrenner's social ecological model [46] and UNICEF conceptual framework for malnutrition [47]. In addition to inadequate dietary intake and disease, food access, household factors, poverty, maternal care practices, health services, the family context, and social values have been implicated in child nutrition outcomes [46,48–50]. Furthermore, nutrition-related knowledge of mothers influenced by living standards and the food environment, as an important component of their care and practices, is necessary although not adequate for optimal healthy child nutrition and positive behavioral change [51,52]. Lastly, data were further informed by the WHO core indicators for optimal IYCF [21,53,54].

2.2. Study Setting and Population

The study was conducted in Mbombela, a local municipality situated in Ehlanzeni District, Mpumalanga Province, described in detail in the previous paper [13]. Mbombela is the capital city of Mpumalanga Province in South Africa with a mixed rural and urban population [55]. In this area, cultural values and norms are highly important. The total population in this district is estimated at 695,913, and there are six community health centers (CHCs) and 31 primary healthcare (PHC) facilities [56]. The six CHCs were chosen based on the fact that they were situated in the rural sections of the settings with poor infrastructure. In such settings, several public health issues such as feeding practices and poor nutritional status are less studied.

This study included children aged under two years and their mothers as a study population. Children were identified during visits to the selected facilities in the study setting for childcare services with their biological mothers and were further eligible to participate in the study after their mothers gave written consent. Children who were brought to these facilities by non-biological mothers were excluded from the study. We further excluded children aged above two years, reported to be sick and on treatment, as well as mothers who were sick with any medical condition that might have affected their feeding practices during the first 6 months prior to giving birth to the participating child, mentally unfit to be interviewed, and below 18 years.

2.3. Sample Size and Sampling Technique

A minimum representative sample size was calculated using total population of 3000 children visiting CHCs [57] and 95% confidence level [58]. A sample size of 405 children paired with their mothers was obtained from the selected CHCs using systematic random sampling. Recruitment of children was conducted when mothers were in the queues for consultation with slips requesting their availability for interviews and anthropometric measurements.

2.4. Data Collection and Tools

2.4.1. Socio-Demographic Status

Data were collected using a validated questionnaire administered by trained research assistants. The questionnaire was adapted from studies conducted with a similar research concept [8,17,18] to collect data on anthropometry, obstetric history, socio-demographic status, feeding practices, and nutrition knowledge from mothers and their children under 2 years. SiSwati-speaking independent translators conversant in English translated the questionnaire forward and backward between the two languages [59]. Content, construct, and face validity were ensured and approved by an expert committee. Following obtaining ethical approval from the Sefako Makgatho Health Sciences University Research and Ethics Committee (SMUREC) (SMUREC/H/290/2020: PG) and permission to access facilities in Mpumalanga Province (MP_202012_005), the study commenced. First, research assistants were trained to collect data, followed by a pilot study conducted among 30 participants to test the feasibility of the study and assess the efficiency of the research assistants to measure anthropometry and administer the questionnaire to participants. A reliability coefficient of 0.82 (i.e., Cronbach's alpha) for the questionnaire was obtained. The results of the pilot study did not form part of the main study but informed improvement of the questionnaire in terms of the wording and structure.

2.4.2. Anthropometric Measurements

Anthropometry of both children and mothers was measured according to WHO procedures [60]. Seca 354 baby (Medicare, Germany) and smart D-quip electronic scales were used to measure weights (i.e., three measurements) of children and mothers to the nearest 10 g and 0.1 kg, respectively. Recumbent length (L) of children and height of mothers were measured using a Seca 210 measuring mat (Medicare, Germany) and stadiometer to the nearest 0.1 cm. Waist and hip circumferences were also measured from mothers using a non-stretchable tape measure to the nearest 0.1 cm. Identity number, date of birth, age, sex, weight, and height of children were captured on the WHO Anthro software version 3.2.2.1 and analyzed according to WHO z-score classification for length-for-age (LAZ), weight-for-age (WAZ), and body mass index-for-age (BAZ) [61]. Stunting (i.e., low LAZ/HAZ), underweight (i.e., low WAZ), and thinness (low BAZ) were defined at <-2SD, while overweight was defined by BAZ $>$ 2SD, and obesity by BAZ $>$ 3SD [60]. The calculated body mass index (BMI) defined underweight ($<$19 kg/m^2), normal weight (19–24.99 kg/m^2), overweight ($\geq$25 kg/m^2), and obesity ($\geq$30 kg/m^2), while abdominal obesity was defined as waist circumference (WC) $\geq$ 88 cm, waist–hip ratio (WHR) $\geq$ 0.85, and waist-to-height ratio (WHtR) $\geq$ 0.5.

2.4.3. Mothers' Feeding Practices and Nutrition-Related Knowledge

Feeding practices include early initiation of breastfeeding, exclusive breastfeeding under 6 months, continued breastfeeding at 1 year, introduction of solid, semi-solid, or soft foods, dietary diversity, minimum meal frequency, minimum acceptable diet, and consumption of iron-rich or iron-fortified foods (eight WHO core indicators of IYCF) [21,53,62].

Adapted from Fadare et al. [30], we used colostrum, continued breastfeeding, diarrhea prevention and treatment using oral rehydration solution, immunization, and family planning to assess mothers' nutrition-related knowledge. Mothers' ability to answer "yes" or "no" to a set of questions relating to child health and nutrition was assessed, and the nutrition-related knowledge was ascertained using frequencies and percentages of the correct, incorrect, or "don't know" answers and categorized into excellent (80–100), good (60–79), average (40–59), and fair (0–39).

2.5. Data Analysis

Descriptive and inferential statistics were computed using STATA version 17 (StataCorp. 2021. Stata Statistical Software: Release 14, College Station, TX, USA). Missing data were assessed using complete case analysis. Distribution of data was assessed using the Shapiro–Francia test. Results are presented as median (interquartile range (IQR)). A chi-square test (or Fisher's exact test for cells with fewer than five cases) was used to compare child nutrition outcomes and selected feeding practices by maternal nutrition-related knowledge. Univariate ($p < 0.20$) and multivariate logistic regression ($p < 0.05$) analyses were used to determine the association of stunting with covariates, controlling for confounding. The main nutritional outcome of interest is child LAZ, which measures long-term nutritional outcomes, and association of stunting and covariates was determined using the crude odds ratio (COR) and adjusted odds ratios (AOR) and significance was considered at $p < 0.05$.

3. Results

3.1. Socio-Demographic Status

3.1.1. Characteristics of Mothers

Information on socio-demographic factors, anthropometry, and obstetric history of mothers is presented in Table 1. The study consisted of mothers participating with their children (n = 400). The mean age of mothers was 29 ± 6 years with the youngest aged 15 years and the oldest aged 42 years. Two age groups were created in this study with 63% of women being younger (<30 years) and 39% being older women. BMI was defined as underweight (<19 kg/m^2), normal weight (19–24.99 kg/m^2), overweight (≥25 kg/m^2), and obesity (≥30 kg/m^2). Mean average BMI of mothers was 28.2 kg/m^2 (±SD = 5) with a range of 19–39 kg/m^2. Most mothers were underweight (39%) or overweight (34%). Abdominal obesity was defined as waist circumference (WC) ≥ 88 cm, waist–hip ratio (WHR) ≥ 0.85, and waist-to-height ratio (WHtR) ≥ 0.5. WC was 72 cm (±SD = 6.3) with a minimum of 64 cm and maximum of 100 cm and WHtR was 0.45 ± 0.37 (range: 0.39–0.60). Abdominal obesity was observed among mothers by WHtR (39%). Most mothers in this study were single (73%) and unemployed (83%) and 40% had attained secondary school education. Thirty-five percent (35%) of women lived in self-headed houses and 45% in parent-headed houses, while 76% lived in larger households (≥5 members). Household income was determined using monthly income from family members in South African rands (ZAR) and converted to US dollars (USD). Most mothers (30%) lived with their children in households with less than USD 5542/month (i.e., ZAR 5000/month) in income, followed by USD 5548–27,710 (i.e., ZAR 5000–ZAR 10,000) (27%). Parity of 1 to 2 was observed in 69% of women, and 30% reported having had obstetric complications.

Table 1. Socio-demography, anthropometry, and obstetric history of mothers.

Variables	Categories	Frequency (n)	Percentage (%)	Median (Range)
Age (years)	<30 ≥30	250 150	63 37	29 (15–42)
Weight (kg)				73.6 (44–119)
Height (cm)				161.5 (150–171)
BMI (Kg/m^2)	Normal Underweight Overweight Obesity	105 155 135 5	26 39 34 1	28.2 (19–39)
WC (cm)	Normal Abdominal obesity	374 26	97 3	72 (64–100)
WHtR	Normal Abdominal obesity	243 157	61 39	0.45 (0.39–0.60)
Marital status	Single Married Divorced	292 56 52	73 14 13	

Table 1. *Cont.*

Variables	Categories	Frequency (*n*)	Percentage (%)	Median (Range)
Education level	No education	72	18	
	Primary school	50	13	
	Secondary school	158	40	
	Completed Grade 12	58	15	
	Tertiary	62	16	
Employment status	Employed	83	21	
	Unemployed	317	79	
Household Head	Self	138	35	
	Spouse	53	13	
	Parents	183	46	
	Grandparents	16	4	
	Relatives	10	3	
Household income (USD/monthly)	<5542	61	15	
	5548–27,710	61	15	
	27,716–55,420	109	27	
	55,426–83,130	93	23	
	>83,130	76	19	
Parity	1–2	277	69	
	≥3	123	31	
Number of children	1–2	253	63	
	≥3	147	37	
Obstetric complications	No	282	70	
	Yes	118	30	

BMI stands for body mass index, WC stands for waist circumference, WHtR stand for weight-to-height ratio, and USD stands for US dollars ($).

3.1.2. Characteristics of Children

Table 2 shows the characteristics of children. Out of 400 children, there were 220 girls and 180 boys (mean age: 8 ± 6 months). Few children (n = 92) were aged 12–23 months, while 153 were aged 6–11 months and 155 aged 0–5 months. Low birth weight (LBW) was observed among 20% of children, while 49% were born at a normal weight, and 88% were born at a clinic (38%) or hospital (50%).

Table 2. Characteristics of children.

Variables	Categories	*n*	%
Sex	Boys	180	45
	Girls	220	55
Age (months)	0–5	155	39
	6–11	153	38
	12–23	92	23
Childbirth weight (g)	<2500	79	20
	≥2500	321	80
Childbirth order	First	91	23
	Middle	195	49
	Last	114	28
Place of birth	Home	47	12
	Clinic	153	38
	Hospital	200	50

g stands for grams.

3.2. Nutritional Outcomes of Children

Comparison of children's nutrition outcomes by sex and age using a chi-square test (or Fisher's exact test for cells with fewer than five cases) is presented in Table 3. WHO z-score classification and definitions for length-for-age (LAZ), weight-for-age (WAZ), and BMI-for-age (BAZ) were used to determine the nutrition outcomes: stunting (LAZ < −2 SD), underweight (WAZ < −2SD), overweight (BAZ > 2SD), and obesity (BAZ > 3SD). The prevalence of stunting (50%) was higher than that of underweight (1%) and thinness (4%) among children. In addition, the prevalence of obesity was high (54%), while overweight (13%) and overweight risk (13%) were also observed. No significant differences in stunting, underweight, overweight, and obesity were observed by sex. Significant differences by age were observed for all the nutritional outcomes among children. Children aged 6–12 months had a higher prevalence of stunting compared to their counterparts, while growth problems and obesity increased with age among children aged ≥12 months.

Table 3. Comparison of children's nutrition outcomes by sex and age.

Variables	All *n* (%)	Boys *n* (%)	Girls *n* (%)	*p*-Value
LAZ				
Normal	111 (28)	50 (28)	61 (28)	
Stunting	202 (50)	90 (50)	112 (51)	0.975
Tallness	87 (22)	40 (22)	47 (21)	
WAZ				
Normal	267 (68)	116 (66)	151 (70)	
Underweight	3 (1)	2 (1)	1 (1)	0.595
Growth problem	121 (31)	57 (33)	64 (29)	
BAZ				
Normal	61 (16)	21 (2)	40 (19)	
Thinness	16 (4)	10 (6)	6 (3)	
Overweight risk	51 (13)	22 (13	29 (13)	0.168
Overweight	50 (13)	27 (15)	23 (11)	
Obesity	213 (54)	95 (54)	118 (55)	
Variables	**0–5 Months *n* (%)**	**6–11 Months *n* (%)**	**≥12 Months *n* (%)**	***p*-Value**
LAZ				
Normal	66 (43)	34 (22)	11 (12)	
Stunting	72 (46)	98 (64)	32 (35)	≤0.0001 *
Tallness	17 (11)	21 (14)	49 (53)	
WAZ				
Normal	128 (88)	120 (79)	19 (21)	
Underweight	2 (1)	1 (1)	0	≤0.0001 *
Growth problem	19 (3)	30 (29)	72 (79)	
BAZ				
Normal	42 (28)	10 (7)	9 (10)	
Thinness	4 (3)	12 (8)	0	
Overweight risk	27 (18)	15 (10)	9 (10)	≤0.0001 *
Overweight	21 (14)	18 (12)	11 (12)	
Obesity	55 (37)	96 (64)	62 (68)	

* Stands for significant difference, *n* stands for number of participants, IQR stands for interquartile range, LAZ stands for length-for-age z-scores, WAZ stands for weight-for-age z-scores, and BAZ stands for body mass index-for-age z-scores.

3.3. Associations of Stunting, Feeding Practices, and Nutrition-Related Knowledge

Mothers' Feeding Practices and Nutrition-Related Knowledge

Child nutrition outcomes (dichotomized) and selected feeding practices are compared by mothers' nutrition-related knowledge using a chi-square test (or Fisher's exact test for cells with fewer than five cases) in Table 4. First, 63% of mothers, initiated breastfeeding within 1 h of delivery and 30% exclusively breastfed. Almost half of mothers (48%) introduced complementary feeding when their children were aged less than 3 months and 70% practiced mixed feeding. Twenty-eight percent (28%) of mothers had a fair nutrition knowledge, while 66% had average knowledge, 6% good knowledge, and none of the mothers had excellent knowledge. Mothers' nutrition knowledge was significantly associated with stunting (p = 0.027) among children.

Table 4. Comparison of child nutrition outcomes and feeding practices by mothers' knowledge.

Variables	All n (%)	Fair NK n (%)	Average NK n (%)	Good NK n (%)	p-Value
Stunting					
No	111 (35)	41 (45)	59 (31)	10 (53)	0.027 *
Yes	202 (65)	50 (55)	130 (69)	9 (47)	
Underweight					
No	267 (98)	76 (100)	163 (98)	16 (100)	0.431
Yes	3 (2)	0	3 (2)	0	
Thinness					
No	62 (79)	19 (83)	36 (78)	6 (86)	0.849
Yes	16 (21)	4 (17)	10 (22)	1 (14)	
Child overweight/obesity					
No	88	19 (19)	35 (15)	6 (26)	0.291
Yes	176	82 (81)	203 (85)	17 (74)	
LBW					
No	321 (80)	85 (79)	205 (81)	20 (83)	0.823
Yes	79 (20)	23 (21)	48 (19)	4 (17)	
Initiation of breastfeeding					
Within 1 h	202 (63)	44 (56)	134 (65)	15 (75)	0.196
>1 h (delayed)	119 (37)	37 (44)	71 (35)	5 (25)	
Exclusive breastfeeding					
No	261 (70)	79 (77)	158 (69)	15 (65)	0.220
Yes	108 (30)	23 (23)	72 (31)	8 (35)	
Mixed feeding					
No	120 (30)	35 (32)	76 (30)	4 (17)	0.360
Yes	277 (70)	73 (68)	175 (70)	19 (83)	
Introduction of complementary foods					
≤3 months	144 (48)	41 (47)	86 (45)	11 (65)	0.306
>3 months	158 (52)	46 (52)	104 (56)	6 (35)	

NK stands for nutrition knowledge, LBW stands for low birth weight, * stands for significant difference.

Table 5 shows the relationship of stunting with independent variables. In univariate logistic regression, stunting was associated with average mother's nutrition-related knowledge, child's age, and education (i.e., completed Grade 12) ($p \leq 0.20$). Multivariate analysis showed significant associations of stunting with mother's nutrition-related knowledge (average: AOR = 1.92, 95%CI: 1.12–3.29), child's age (6–11 months: AOR = 2.63, 95%CI: 1.53–4.53 and 12–23 months: AOR = 3.19, 95%CI: 1.41–7.25), and education (completing Grade 12: AOR = 0.36, 95%CI: 0.15–0.86). Therefore, average mother's nutrition-related knowledge had a likelihood to predispose children to stunting, while the odds of stunting were high for children aged 6–11 and 12–23 months compared to those aged 0–5 months and completing grade 12 among mothers was protective against stunting in children compared to having no education.

Table 5. Association of stunting with covariates.

Stunting	COR (95%CI)	*p*-Value	AOR (95%CI)	*p*-Value
Mother's nutrition-related knowledge				
Fair	1		1	
Average	1.81 (1.08–3.02)	0.024 *	1.92 (1.12–3.29)	0.018 *
Good	0.74 (0.27–1.99)	0.548	0.63 (0.22–1.85)	0.403
Child's age (months)				
0–5	1		1	
6–11	2.64 (1.58–4.42)	≤0.0001 *	2.63 (1.53–4.53)	≤0.0001 *
12–23	2.67 (1.24–5.71)	0.012 *	3.19 (1.41–7.25)	0.006 *
Mother's education				
No education	1		1	
Primary	0.73 (0.31–1.74)	0.475	0.90 (0.36–2.27)	0.831
Secondary	0.95 (0.48–1.86)	0.872	1.11 (0.54–2.28)	0.783
Completed Grade 12	0.36 (0.16–0.80)	0.013 *	0.36 (0.15–0.86)	0.021 *
Tertiary	0.79 (0.34–1.85)	0.592	0.75 (0.30–1.85)	0.528

COR stands for crude odds ratio, AOR stands for adjusted odds ratio, and * stand for $p < 0.05$.

4. Discussion

This study reports a high prevalence of stunting (50%) among children, accompanied by an alarming rate of obesity (54%). South Africa continues to experience persistent stunting affecting 22–55% of children [8,11,17,18,38,40], like in other African countries such as Ghana [29], Ethiopia [63], and Nigeria [30,63]. In addition to stunting predisposing children to poor cognition and educational performance as they advance in age [6], impaired fat oxidation associated with stunting makes children susceptible to obesity, including later in life [64,65]. Furthermore, a high prevalence of obesity among children in this study is consistent with other reports in South Africa [11,40,66]. Surprisingly, most women in this study were underweight (39%), like in Ethiopia (30%), Uganda (12%), and Tanzania (11%) [67], where high prevalence of underweight have been reported, which could be multifactorial. The prevalence of overweight (34%) reported in this study contributes to public health concerns in South Africa, although it is lower compared to maternal overweight/obesity in South Africa other studies, estimated at 54–76% [68–70]. High consumption of energy-dense food and reduced physical activity are always implicated in overweight/obesity in other countries [71–73], and in South Africa [41,68], nutrition transition is associated with frequent intake of processed and sugary foods as well as high-calorie and high-fat diets, socio-demographic characteristics, etc. [74–77].

Researchers have implicated chronic nutritional deprivation in stunting [78], usually associated with poor socio-economic conditions, poor maternal health and nutrition, frequent illness, and/or inappropriate infant and young child feeding and care in early life [79]. We attribute stunting in this study to household poverty, as indicated by the poor socio-economic status. Further, feeding practices, especially breastfeeding, are associated with socio-economic status, and in the current study poor feeding practices are characterized by a low level of exclusive breastfeeding and an early introduction of solid foods like in other studies in Africa [80–82] and locally [17,36,83]. On the other hand, children who consume formula milk become overweight/obese due to high protein and energy [66,67]. Furthermore, children who are susceptible to weight gain in the first week of life have a higher risk of weight gain between one and two years, and this doubles as they grow older [39,84,85]. During the first 6 months of life, exclusive breastfeeding protects children against overweight and obesity, among other benefits [13,48,86]. We attribute obesity in this case to poor feeding practices, such as early introduction of solid foods.

In this study, mothers and their children lived in socio-economically poor households and reported poor obstetric history, as also reported in similar South African studies [17,80]. Socio-economic status is one of the most important factors associated with health and medical outcomes of women of reproductive age [81,82]. In addition, women with lower socio-economic status are less likely to receive prenatal care, which is associated with poor obstetric outcomes [81–83]. Socio-economic differences across countries play a significant role [87] and are known to affect the prevalence of the double burden of malnutrition in LMICs, in addition to nutrition transition [42,43,88,89]. Therefore, the coexistence of maternal underweight and obesity, as well as child stunting and obesity, in poorer households suggests a household double burden of malnutrition [41,42,89]. Health implications of the double burden of malnutrition include increased mortality and morbidity risks, as well as poor cognition, health, and development with adult health risks [1,90].

The study has reported increased odds of stunting in children aged 6–11 months and 12–23 months. Increased odds of stunting with age have been reported among children in parts of Ethiopia [91], Indonesia [92], and Zambia [93]. Additionally, this study found average nutrition-related knowledge (66%) among mothers, similar to the 61.5% reported among caregivers in Ghana [94] and 59% in Kenya [95]. However, this knowledge does not seem to be reflected in practice since low levels of exclusive breastfeeding and early introduction of solids were observed, although two thirds were able to initiate breastfeeding within one hour in this study. It could be that in settings like our setting, a low level of parental education in terms of tertiary education might have no significant effect on child nutrition outcomes [60,96]. This kind of education status is unlikely to reinforce a mother's nutrition knowledge for the purposes of reducing poor child nutrition outcomes.

On the contrary, high nutritional knowledge has been reported to be essential to promote good feeding practices and prevent poor nutrition outcomes among children in Kenya [97]. Our study reported a positive relationship between stunting and the mother's education level, the same as in the study of Negash et al. [98] in Ethiopia and Fadare et al. [30] in Nigeria. A higher chance that a mother who completed Grade 12 would have a child who is less likely to be stunted compared to a mother with no education was observed in this study. It has been suggested that mothers with above a primary-level education can significantly reduce stunting in children [30]. Furthermore, the odds of mothers with average nutrition-related knowledge to have stunted children were 1.92 [29]. In Ghana, a positive link between a caregiver's nutritional knowledge and stunting was reported among children aged 6–36 months [99]. Another study in Ghana, documented the effect of nutritional knowledge of mothers on children's nutritional status and child feeding knowledge and practices were associated positively with appropriate feeding practices [96].

The significant difference in feeding practices of children in different stages of growth is well documented, and this requires mothers to have diverse nutrition knowledge to cater for those stages [30,100]. In this study, mothers were expected to recall their feeding practices from the outset after birth to the time the study was conducted. From our results, it is shown that the nutrition knowledge of mothers was not significantly different by children's age group for the phases of exclusive breastfeeding in the first 6 months, introduction to solid food, as well as mixed feeding. This suggests a needed change in nutrition knowledge of mothers to feed their children appropriately as they advance in the various stages of growth, because their knowledge of food choices, feeding, and healthcare seeking is vital for producing normal child nutrition outcomes. For instance, as infants age beyond six months, they become more vulnerable to poor nutrition outcomes during the transition period from a milk diet to a diet that includes complementary food, predisposing them to stunting. This nutritional outcome is worsened by continued breastfeeding not accompanied by adequate complementary feeding at the appropriate age, due to nutritional demand [19–21].

Lastly, the context in which children grow up in can never be ignored as one of the influences on their nutrition outcomes, and the literature documents family context, ethnicity, and social values [12,101,102]. The impact of cultural beliefs and traditional practices is acknowledged in this study considering that the study setting is known for strong cultural values and norms, as well as poverty. First, the context of poverty in the current study, explained by poor living conditions, predisposes households to food insecurity through limited availability, access, and affordability and this will ultimately prevent children from meeting their nutrient requirements for growth [12]. Second, cultural beliefs and traditional practices contribute to feeding practices and vary across communities and countries [10,103–105]. Like in other countries, the role of mothers/caregivers regarding childcare is highly emphasized in most traditional cultures, where intergenerational influences on meal preparation inform systems and practices [105,106]. For instance, over two thirds of children in this study were non-exclusively breastfed and mixed fed. These are normalized feeding practices in other African societies with health impacts on children's health due to exposure to the risk of diarrhea and malnutrition [5,102,107–111]. Furthermore, the pressure exerted by the family to give the baby other food further exacerbates the problem of early introduction of solids and other liquids [108,110], which might explain why almost half of the children in this study started to consume solid foods under 3 months of age. However, other researchers have acknowledged that some traditional practices, while often times misleading and not scientifically always sound, have been shown under certain circumstances to actually improve health/nutritional outcomes in children because they capitalize optimally on limited resources [105]. Therefore, in settings such as ours, challenges remain on how to assist mothers to make informed decisions regarding IYCF practices, considering culturally appropriate infant feeding messages.

Strengths and Limitations of the Study

Inferences made in this cross-sectional study are reliable, with limitations on establishing causality using this type of study design. We were able to obtain a random sampling, which mitigated selection bias. However, when it comes to collecting information from mothers, social desirability is the most concerning limitation since mothers might have withheld information or falsified their responses, especially on feeding practices, with the aim to be acceptable. We relied on self-reported information mostly, and having to remember previous events and practices might have also introduced recall biases. We cannot explain the high prevalence of underweight among women, as South Africa is known to have minimal levels of underweight among women of reproductive age and high levels of overweight/obesity compared to their counterparts. Conducting a health assessment of mothers including medical history might have provided a better explanation of underweight among mothers. Further, care should be taken when determining feeding practices; nonetheless, we used a WHO questionnaire, which was piloted prior to commencing the main study and entailed training research assistants on conducting interviews and anthropometry. Lastly, we did not measure the HIV status of mothers, which could have provided a better indication regarding their feeding practices, especially regarding breastfeeding. Also, confounding factors such as medication use and smoking and alcohol use as lifestyle factors should be considered in future studies. The results of this study are applicable to rural localities in Mbombela, Mpumalanga Province in South Africa and cannot be generalized to other areas because setting variations may influence feeding practices. Nevertheless, this study has reported on poor child nutrition outcomes, feeding practices, mothers' nutritional knowledge, and their relationship.

5. Conclusions

The high prevalence of stunting and obesity among children found in this study has great policy implications for South Africa, considering the lack of improved child nutrition outcomes, despite the introduction of the Integrated Nutrition Programme. Furthermore, poor feeding practices remains in South Africa even after having promulgated several policies regarding feeding of children. At the same time, poor maternal nutrition-related knowledge was prevalent, but most mothers had average nutrition knowledge.

These findings contribute significant information for understanding the relationship between maternal nutrition-related knowledge and child nutrition outcomes in the context of a poverty-stricken environment. A significant association of stunting and mothers' nutrition-related knowledge might be helpful in reducing stunting occurrences; however, this knowledge might not reflect good feeding practice. In addition, the association of mothers completing Grade 12 with lower odds of stunting among children could suggest that a low level of parental education in terms of tertiary education might not likely reinforce mothers' nutrition knowledge for the purposes of reducing poor child nutrition outcomes. Therefore, contextual, and intensified interventions on continued education for mothers to gain accurate information on nutrition and feeding practices could ultimately enhance child nutrition outcomes in poorer settings. Efforts should therefore be made to ensure that nutrition knowledge is appropriately provided based on the phases of child growth from 0 to 2 years, even beyond infancy into school age.

Author Contributions: Conceptualization, L.N.M. and P.M.; methodology, L.N.M. and P.M.; software, P.M.; validation, L.N.M.; formal analysis, P.M.; investigation, L.N.M.; resources, L.N.M.; data curation, L.N.M.; and P.M.; writing—original draft preparation, P.M.; writing—review and editing, L.N.M. and P.M.; supervision, P.M.; project administration, L.N.M. All authors have read and agreed to the published version of the manuscript.

Funding: This research received no external funding.

Institutional Review Board Statement: This study was conducted according to the guidelines laid down in the Declaration of Helsinki [112] and all procedures involving human subjects were approved by Sefako Makgatho Health Sciences University Research and Ethics Committee (SMUREC) (SMUREC/H/290/2020: PG), approved on 1 June 2017. The study further received permission, from the Provincial Department of Health Research Committee, in Mpumalanga Province of South Africa (MP_202012_005), on 21 December 2020.

Informed Consent Statement: Mothers attending childcare services received information about the study. Following recruitment and information distribution, mothers gave written informed consent to participate in the study with their children.

Data Availability Statement: The dataset for the study groups generated and analyzed during the current study is available from the corresponding author upon reasonable request due to ethical restrictions.

Acknowledgments: We thank mothers for willingly participating in this study with their children. We acknowledge the Mpumalanga Provincial Department of Health Committee, in South Africa. We are grateful to the facility managers for granting us access to conduct this research and further assisting us with recruitment during visits to the facilities, as well as the research assistants for data collection.

Conflicts of Interest: The authors declare no conflict of interest.

References

1. De Onis, M.; Blössner, M.; Borghi, E.; Morris, R.; Frongillo, E.A. Methodology for estimating regional and global trends of child malnutrition. *Int. J. Epidemiol.* **2004**, *33*, 1260–1270. [CrossRef]
2. Quamme, S.H.; Iversen, P.O. Prevalence of child stunting in Sub-Saharan Africa and its risk factors. *Clin. Nutr. Open Sci.* **2022**, *42*, 49–61. [CrossRef]
3. UNICEF. Reduce Stunting. Chronic Malnutrition Will Result in Stunting—An Irreversible Condition That Literally Stunts the Physical and Cognitive Growth of Children. 2022. Available online: https://www.unicef.org/esa/reduce-stunting (accessed on 15 August 2022).
4. Dewey, K.G.; Begum, K. Long-term consequences of stunting in early life. *Matern. Child Nutr.* **2011**, *7* (Suppl. S3), 5–18. [CrossRef]
5. Martorell, R.; Melgar, P.; Maluccio, J.A.; Stein, A.D.; Rivera, J.A. The nutrition intervention improved adult human capital and economic productivity. *J. Nutr.* **2010**, *140*, 411–414. [CrossRef] [PubMed]
6. de Onis, M.; Branca, F. Childhood stunting: A global perspective. *Matern. Child Nutr.* **2016**, *12* (Suppl. S1), 12–26. [CrossRef] [PubMed]
7. Devkota, M.D.; Adhikari, R.K.; Upreti, S.R. Stunting in Nepal: Looking back, looking ahead. *Matern. Child Nutr.* **2016**, *12* (Suppl. S1), 257–259. [CrossRef]

8. Symington, E.A.; Gericke, G.J.; Nel, J.H.; Labadarios, D. The relationship between stunting and overweight among children from South Africa: Secondary analysis of the National Food Consumption Survey—Fortification Baseline I. *S. Afr. Med. J.* **2015**, *106*, 65–69. [CrossRef]
9. World Health Organization. *Global Nutrition Targets 2025: Stunting Policy Brief*; World Health Organization: Geneva, Switzerland, 2014.
10. Modjadji, P.; Madiba, S. The Multidimension of Malnutrition among School Children in a Rural Area, South Africa: A Mixed Methods Approach. *Nutrients* **2022**, *14*, 5015.
11. NDoH; Stats SA; SAMRC; ICF. *South Africa Demographic and Health Survey 2016: Key Indicators*; Statistics South Africa: Pretoria, South Africa; Rockville, MD, USA, 2017.
12. Modjadji, P. Engaging Mothers on the Growth of School-Age Children in a Rural South African Health and Demographic Site: A Qualitative Insight. *Healthcare* **2021**, *9*, 225.
13. Modjadji, P.; Masilela, L.N.; Cele, L.; Mathibe, M.; Mphekgwana, P.M. Evidence of Concurrent Stunting and Obesity among Children under 2 Years from Socio-Economically Disadvantaged Backgrounds in the Era of the Integrated Nutrition Programme in South Africa. *Int. J. Environ. Res. Public Health* **2022**, *19*, 12501. [CrossRef]
14. Amaha, N.D.; Woldeamanuel, B.T. Maternal factors associated with moderate and severe stunting in Ethiopian children: Analysis of some environmental factors based on 2016 demographic health survey. *Nutr. J.* **2021**, *20*, 18. [CrossRef] [PubMed]
15. Das, N.; Chattopadhyay, D.; Chakraborty, S.; Dasgupta, A. Infant and young child feeding perceptions and practices among mothers in a rural area of West Bengal, India. *Ann. Med. Health Sci. Res.* **2013**, *3*, 370–375. [PubMed]
16. Sanghvi, T.; Martin, L.; Hajeebhoy, N.; Abrha, T.H.; Abebe, Y.; Haque, R.; Tran, H.T.T.; Roy, S. Strengthening systems to support mothers in infant and young child feeding at scale. *Food Nutr. Bull.* **2013**, *34*, S156–S168. [CrossRef] [PubMed]
17. Modjadji, P.; Mashishi, J. Persistent Malnutrition and Associated Factors among Children under Five Years Attending Primary Health Care Facilities in Limpopo Province, South Africa. *Int. J. Environ. Res. Public Health* **2020**, *17*, 7580. [CrossRef]
18. Modjadji, P.; Pitso, M. Maternal Tobacco and Alcohol Use in Relation to Child Malnutrition in Gauteng, South Africa: A Retrospective Analysis. *Children* **2021**, *8*, 133. [CrossRef]
19. Thiombiano-Coulibaly, N.; Rocquelin, G.; Eymard-Duvernay, S.; Zougmore, O.; Traore, S. Effects of early extra fluid and food intake on breast milk consumption and infant nutritional status at 5 months of age in an urban and a rural area of Burkina Faso. *Eur. J. Clin. Nutr.* **2004**, *58*, 80–89. [CrossRef]
20. Derso, T.; Tariku, A.; Biks, G.A.; Wassie, M.M. Stunting, wasting and associated factors among children aged 6–24 months in Dabat health and demographic surveillance system site: A community based cross-sectional study in Ethiopia. *BMC Pediatr.* **2017**, *17*, 96. [CrossRef]
21. Black, R.E.; Allen, L.H.; Bhutta, Z.A.; Caulfield, L.E.; de Onis, M.; Ezzati, M.; Mathers, C.; Rivera, J. Maternal and child undernutrition: Global and regional exposures and health consequences. *Lancet* **2008**, *371*, 243–260. [CrossRef]
22. Saito, K.; Korzenik, J.R.; Jekel, J.F.; Bhattacharji, S. A case-control study of maternal knowledge of malnutrition and health-care-seeking attitudes in rural South India. *Yale J. Biol. Med.* **1997**, *70*, 149–160. Available online: https://pubmed.ncbi.nlm.nih.gov/9493847/ (accessed on 29 June 2023).
23. Haddad, L.; Achadi, E.; Bendech, M.A.; Ahuja, A.; Bhatia, K.; Bhutta, Z.; Blössner, M.; Borghi, E.; Colecraft, E.; De Onis, M. The Global Nutrition Report 2014: Actions and accountability to accelerate the world's progress on nutrition. *J. Nutr.* **2015**, *145*, 663–671. [CrossRef]
24. Sayed, N.; Schönfeldt, H.C. A review of complementary feeding practices in South Africa. *S. Afr. J. Clin. Nutr.* **2020**, *33*, 36–43. [CrossRef]
25. UNICEF. *Strategy for Improved Nutrition of Children and Woman in Developing Counties*; UNICEF: New York, NY, USA, 1990.
26. Bengre, A. Mothers knowledge on Malnutrition: Community Based Cross Sectional Study. *Indian J. Public Health Res. Dev.* **2018**, *9*, 37–41. [CrossRef]
27. Glewwe, P. Why does mother's schooling raise child health in developing countries? Evidence from Morocco. *J. Hum. Resour.* **1999**, *34*, 124–159. [CrossRef]
28. Gittelsohn, J.; Vastine, A.E. Sociocultural and household factors impacting on the selection, allocation and consumption of animal source foods: Current knowledge and application. *J. Nutr.* **2003**, *133*, 4036S–4041S. [CrossRef] [PubMed]
29. Appoh, L.Y.; Krekling, S. Maternal nutritional knowledge and child nutritional status in the Volta region of Ghana. *Matern. Child Nutr.* **2005**, *1*, 100–110. [CrossRef]
30. Fadare, O.; Amare, M.; Mavrotas, G.; Akerele, D.; Ogunniyi, A. Mother's nutrition-related knowledge and child nutrition outcomes: Empirical evidence from Nigeria. *PLoS ONE* **2019**, *14*, e0212775. [CrossRef]
31. Motadi, S.; Malise, T.; Mushaphi, L. Breastfeeding knowledge and practices among mothers of children younger than 2 years from a rural area in the Limpopo province, South Africa. *S. Afr. J. Child. Health* **2019**, *13*, 115. [CrossRef]
32. Webb, P.; Block, S.A. Nutrition Knowledge and Parental Schooling as Inputs to Child Nutrition in the Long and Short Run. Tufts University Friedman School of Nutrition Science and Policy. Nutrition Working Paper 3. 2003. Available online: https://www.researchgate.net/profile/Steven-Block-4/publication/5057561_Nutrition_Knowledge_and_Parental_Schooling_as_Inputs_to_Child_Nutrition_in_the_Long_and_Short_Run/links/0fcfd50e445abb37b3000000/Nutrition-Knowledge-and-Parental-Schooling-as-Inputs-to-Child-Nutrition-in-the-Long-and-Short-Run.pdf (accessed on 29 June 2023).

33. Novitasari, P.; Wanda, D. Maternal Feeding Practice and Its Relationship with Stunting in Children. *Pediatr. Rep* **2020**, *12*, 8698. [CrossRef] [PubMed]
34. Claeson, M.; Waldman, R.J. The evolution of child health programmes in developing countries: From targeting diseases to targeting people. *Bull. World Health Organ.* **2000**, *78*, 1234–1245. Available online: https://pubmed.ncbi.nlm.nih.gov/11100618/ (accessed on 29 June 2023). [PubMed]
35. Lee, R.G.; Garvin, T. Moving from information transfer to information exchange in health and health care. *Soc. Sci Med.* **2003**, *56*, 449–464. [CrossRef]
36. Chakona, G. Social circumstances and cultural beliefs influence maternal nutrition, breastfeeding and child feeding practices in South Africa. *Nutr. J.* **2020**, *19*, 47. [CrossRef]
37. Mushaphi, L.; Mbhenyane, X.; Khoza, L.; Amey, A. Infant-feeding practices of mothers and the nutritional status of infants in the Vhembe District of Limpopo Province. *S. Afr. J. Clin. Nutr.* **2008**, *21*, 36–41. [CrossRef]
38. Modjadji, P.; Molokwane, D.; Ukegbu, P.O. Dietary Diversity and Nutritional Status of Preschool Children in North West Province, South Africa: A Cross Sectional Study. *Children* **2020**, *7*, 174. [CrossRef]
39. Mamabolo, R.L.; Steyn, N.P.; Alberts, M. Can the high prevalence of micronutrient deficiencies, stunting and overweight in children at ages 1 and 3 years in the Central Region of Limpopo province be explained by diet? *S. Afr. J. Clin. Nutr.* **2006**, *19*, 102–113. Available online: http://www.sajcn.co.za/index.php/SAJCN/article/view/146 (accessed on 29 June 2023). [CrossRef]
40. Mamabolo, R.L.; Alberts, M.; Steyn, N.P.; Delemarre-van de Waal, H.A.; Levitt, N.S. Prevalence and determinants of stunting and overweight in 3-year-old black South African children residing in the Central Region of Limpopo Province, South Africa. *Public Health Nutr.* **2005**, *8*, 501–508. [CrossRef] [PubMed]
41. Modjadji, P.; Madiba, S. The double burden of malnutrition in a rural health and demographic surveillance system site in South Africa: A study of primary schoolchildren and their mothers. *BMC Public Health* **2019**, *19*, 1087. [CrossRef]
42. Kimani-Murage, E.W. Exploring the paradox: Double burden of malnutrition in rural South Africa. *Glob. Health Action* **2013**, *6*, 19249. [CrossRef] [PubMed]
43. Kimani-Murage, E.W.; Kahn, K.; Pettifor, J.M.; Tollman, S.M.; Dunger, D.B.; Gómez-Olivé, X.F.; Norris, S.A. The prevalence of stunting, overweight and obesity, and metabolic disease risk in rural South African children. *BMC Public Health* **2010**, *10*, 158. [CrossRef]
44. Seonandan, P.; McKerrow, N.H. A review of infant and young child feeding practice in hospital and the home in KwaZulu-Natal Midlands. *S. Afr. J. Clin. Nutr.* **2016**, *29*, 111–115. [CrossRef]
45. Mwangome, M.; Prentice, A.; Plugge, E.; Nweneka, C. Determinants of appropriate child health and nutrition practices among women in rural Gambia. *J. Health Popul. Nutr.* **2010**, *28*, 167–172. [CrossRef]
46. Bronfenbrenner, U. Ecological models of human development. *Int. Encycl. Educ.* **1994**, *3*, 37–43.
47. UNICEF. Conceptual frameworks. In *UNICEF M&E Training Resources*; Oxford University Press: New York, NY, USA, 1998.
48. Bronfenbrenner, U. Ecology of the family as a context for human development: Research perspectives. *Dev. Psychol.* **1986**, *22*, 723–724. [CrossRef]
49. Darling, N.; Steinberg, L. Parenting style as context: An integrative model. *Psychol. Bull.* **1993**, *113*, 487. [CrossRef]
50. Pinquart, M.; Kauser, R. Do the associations of parenting styles with behavior problems and academic achievement vary by culture? Results from a meta-analysis. *Cult. Divers. Ethn. Minor. Psychol.* **2018**, *24*, 75. [CrossRef] [PubMed]
51. Contento, I.R. Nutrition education: Linking research, theory, and practice. *Asia Pac. J. Clin. Nutr.* **2008**, *17* (Suppl. 1), 176–179. Available online: https://pubmed.ncbi.nlm.nih.gov/18296331/ (accessed on 29 June 2023). [PubMed]
52. Hawkes, C.; Smith, T.G.; Jewell, J.; Wardle, J.; Hammond, R.A.; Friel, S.; Thow, A.M.; Kain, J. Smart food policies for obesity prevention. *Lancet* **2015**, *385*, 2410–2421. [CrossRef]
53. Miah, R.; Apanga, P.; Abdul-haq, Z. Risk Factors for Undernutrition in Children under Five Years Old: Evidence from the 2011 Ghana Multiple Indicator Cluster Survey. *J. AIDS Clin. Res.* **2016**, *7*, 585. [CrossRef]
54. WHO Multicentre Growth Reference Study Group. Complementary feeding in the WHO multicentre growth reference study. *Acta Paediatr.* **2006**, *450*, 27–37.
55. Mpumalanga Provincial Government. Municipalities of South Africa; Ehlanzeni District Municipality (DC32). 2012–2022. Available online: https://municipalities.co.za/overview/131/ehlanzeni-district-municipality (accessed on 2 February 2022).
56. Drigo, L.; Luvhengo, M.; Lebese, R.T.; Makhado, L. Attitudes of Pregnant Women towards Antenatal Care Services Provided in Primary Health Care Facilities of Mbombela Municipality, Mpumalanga Province, South Africa. *Open Public Health J.* **2020**, *13*, 569–575. [CrossRef]
57. South Africa Web DHIS. Available online: https://za.dhis.dhmis.org/dhis-web-pivot/index.html (accessed on 23 February 2021).
58. Raosoft. Raosoft Sample Size Calculator. 2004. Available online: http://www.raosoft.com/samplesize.html (accessed on 2 February 2022).
59. Tsang, S.; Royse, C.F.; Terkawi, A.S. Guidelines for developing, translating, and validating a questionnaire in perioperative and pain medicine. *Saudi J. Anaesth.* **2017**, *11*, S80. [CrossRef]
60. World Health Organization. *Guideline: Assessing and Managing Children at Primary Health-Care Facilities to Prevent Overweight and Obesity in the Context of the Double Burden of Malnutrition*; World Health Organization: Geneva, Switzerland, 2017.
61. World Health Organization. *WHO Anthro for Personal Computers, Version 3.2. 2, 2011: Software for Assessing Growth and Development of the World's Children*; World Health Organization: Geneva, Switzerland, 2010.

62. World Health Organization. *Indicators for Assessing Infant and Young Child Feeding Practices: Part 1: Definitions: Conclusions of a Consensus Meeting Held 6–8 November 2007 in Washington DC, USA*; World Health Organization: Geneva, Switzerland, 2008.
63. Amare, M.; Benson, T.; Fadare, O.; Oyeyemi, M. Study of the Determinants of Chronic Malnutrition in Northern Nigeria: Quantitative Evidence from the Nigeria Demographic and Health Surveys International Food Policy Research Institute (IFPRI) Working Paper 45 (September 2017). *Food Nutr. Bull.* **2018**, *39*, 296–314. [CrossRef] [PubMed]
64. Hoffman, D.J.; Sawaya, A.L.; Verreschi, I.; Tucker, K.L.; Roberts, S.B. Why are nutritionally stunted children at increased risk of obesity? Studies of metabolic rate and fat oxidation in shantytown children from Sao Paulo, Brazil. *Am. J. Clin. Nutr.* **2000**, *72*, 702–707. [CrossRef] [PubMed]
65. Gribble, J.N.; Murray, N.J.; Menotti, E.P. Reconsidering childhood undernutrition: Can birth spacing make a difference? An analysis of the 2002-2003 El Salvador National Family Health Survey. *Matern. Child Nutr.* **2009**, *5*, 49–63. [CrossRef] [PubMed]
66. Tathiah, N.; Moodley, I.; Mubaiwa, V.; Denny, L.; Taylor, M. South Africa's nutritional transition: Overweight, obesity, underweight and stunting in female primary school learners in rural KwaZulu-Natal, South Africa. *SAMJ S. Afr. Med. J.* **2013**, *103*, 718–723. [CrossRef] [PubMed]
67. Mtumwa, A.H.; Paul, E.; Vuai, S.A.H. Determinants of undernutrition among women of reproductive age in Tanzania mainland. *S. Afr. J. Clin. Nutr.* **2016**, *29*, 75–81. [CrossRef]
68. Maimela, E.; Alberts, M.; Modjadji, S.E.P.; Choma, S.S.R.; Dikotope, S.A.; Ntuli, T.S.; Van Geertruyden, J.-P. The Prevalence and Determinants of Chronic Non-Communicable Disease Risk Factors amongst Adults in the Dikgale Health Demographic and Surveillance System (HDSS) Site, Limpopo Province of South Africa. *PLoS ONE* **2016**, *11*, e0147926. [CrossRef]
69. Mashinya, F.; Alberts, M.; Cook, I.; Ntuli, S. Determinants of body mass index by gender in the Dikgale Health and Demographic Surveillance System site, South Africa. *Glob. Health Action* **2018**, *11*, 1537613. [CrossRef]
70. Tydeman-Edwards, R. Obsesity, Undernutrition and the Double Burden of Disease in the Free State. Ph.D. Thesis, University of the Free State, Bloemfontein, South Africa, 2012.
71. Alemu, F.M. Assessment of the impact of malnutrition on children at Dilla referral hospital and unity pediatric clinic, Ethiopia. *Int. J. Nutr. Metab.* **2013**, *5*, 105–113. [CrossRef]
72. Dieffenbach, S.; Stein, A.D. Stunted child/overweight mother pairs represent a statistical artifact, not a distinct entity. *J. Nutr.* **2012**, *142*, 771–773. [CrossRef]
73. Guthold, R.; Ono, T.; Strong, K.L.; Chatterji, S.; Morabia, A. Worldwide variability in physical inactivity a 51-country survey. *Am. J. Prev. Med.* **2008**, *34*, 486–494. [CrossRef]
74. Adienbo, O. High Prevalence of Obesity among Indigenous Residents of a Nigerian Ethnic Group: The Kalabaris in the Niger Delta Region of South-South Nigeria. *Greener J. Med. Sci.* **2012**, *2*, 152–156. [CrossRef]
75. Mukora-Mutseyekwa, F.; Zeeb, H.; Nengomasha, L.; Kofi Adjei, N. Trends in Prevalence and Related Risk Factors of Overweight and Obesity among Women of Reproductive Age in Zimbabwe, 2005–2015. *Int. J. Environ. Res. Public Health* **2019**, *16*, 2758. [CrossRef] [PubMed]
76. Rothman, M.; Ranneileng, M.; Nel, R.; Walsh, C. Nutritional status and food intake of women residing in rural and urban areas of Lesotho. *S. Afr. J. Clin. Nutr.* **2019**, *32*, 21–27. [CrossRef]
77. Steyn, N.P.; Labadarios, D.; Nel, J.; Kruger, H.S.; Maunder, E.M.W. What is the nutritional status of children of obese mothers in South Africa? *Nutrition* **2011**, *27*, 904–911. [CrossRef]
78. Perkins, J.M.; Kim, R.; Krishna, A.; McGovern, M.; Aguayo, V.M.; Subramanian, S.V. Understanding the association between stunting and child development in low- and middle-income countries: Next steps for research and intervention. *Soc. Sci. Med. (1982)* **2017**, *193*, 101–109. [CrossRef]
79. WHO. *Malnutrition*; Whorld Health Organization: Geneva, Switzerland, 2021; Available online: https://www.who.int/news-room/fact-sheets/detail/malnutrition#:~:text=Low%20height-for-age%20is%20known%20as%20stunting.%20It%20is,young%20child%20feeding%20and%20care%20in%20early%20life (accessed on 27 June 2023).
80. Goon, D.T.; Ajayi, A.I.; Adeniyi, O.V. Sociodemographic and lifestyle correlates of exclusive breastfeeding practices among mothers on antiretroviral therapy in the Eastern Cape, South Africa. *Int. Breastfeed. J.* **2021**, *16*, 18. [CrossRef]
81. Lee, S.H.; Lee, S.M.; Lim, N.G.; Kim, H.J.; Bae, S.H.; Ock, M.; Kim, U.N.; Lee, J.Y.; Jo, M.W. Differences in pregnancy outcomes, prenatal care utilization, and maternal complications between teenagers and adult women in Korea: A nationwide epidemiological study. *Medicine* **2016**, *95*, e4630. [CrossRef] [PubMed]
82. Riva, E.; Banderali, G.; Agostoni, C.; Silano, M.; Radaelli, G.; Giovannini, M. Factors associated with initiation and duration of breastfeeding in Italy. *Acta Paediatr.* **1999**, *88*, 411–415. [CrossRef]
83. Pei, L.; Kang, Y.; Zhao, Y.; Cheng, Y.; Yan, H. Changes in Socioeconomic Inequality of Low Birth Weight and Macrosomia in Shaanxi Province of Northwest China, 2010–2013: A Cross-sectional Study. *Medicine* **2016**, *95*, e2471. [CrossRef]
84. Koplan, J.P.; Liverman, C.T.; Kraak, V.I. *Preventing Childhood Obesity: Health in the Balance*; National Academies Press (US): Washington, DC, USA, 2005. Available online: https://pubmed.ncbi.nlm.nih.gov/22379642/ (accessed on 14 June 2023).
85. Ong, K.K.; Loos, R.J. Rapid infancy weight gain and subsequent obesity: Systematic reviews and hopeful suggestions. *Acta Paediatr.* **2006**, *95*, 904–908. [CrossRef]
86. Stewart, C.P.; Iannotti, L.; Dewey, K.G.; Michaelsen, K.F.; Onyango, A.W. Contextualising complementary feeding in a broader framework for stunting prevention. *Matern. Child Nutr.* **2013**, *9*, 27–45. [CrossRef]

87. Garrett, J.L.; Ruel, M.T. Stunted child-overweight mother pairs: Prevalence and association with economic development and urbanization. *Food Nutr. Bull.* **2005**, *26*, 209–221. [CrossRef]
88. Min, J.; Zhao, Y.; Slivka, L.; Wang, Y. Double burden of diseases worldwide: Coexistence of undernutrition and overnutrition-related non-communicable chronic diseases. *Obes. Rev. Off. J. Int. Assoc. Study Obes.* **2018**, *19*, 49–61. [CrossRef] [PubMed]
89. Popkin, B.M. The shift in stages of the nutrition transition in the developing world differs from past experiences! *Public Health Nutr.* **2002**, *5*, 205–214. [CrossRef]
90. De Haas, H. Migration and development: A theoretical perspective. *Int. Migr. Rev.* **2010**, *44*, 227–264. [CrossRef]
91. Lisanu Mazengia, A.; Andargie Biks, G. Predictors of Stunting among School-Age Children in Northwestern Ethiopia. *J. Nutr. Metab.* **2018**, *2018*, 7521751. [CrossRef] [PubMed]
92. Yasmin, G.; Kustiyah, L.; Dwiriani, C.M. Risk factors of stunting among school-aged children from eight provinces in Indonesia. *Pak. J. Nutr.* **2014**, *13*, 557. [CrossRef]
93. Mzumara, B.; Bwembya, P.; Halwiindi, H.; Mugode, R.; Banda, J. Factors associated with stunting among children below five years of age in Zambia: Evidence from the 2014 Zambia demographic and health survey. *BMC Nutr.* **2018**, *4*, 51. [CrossRef]
94. Forh, G.; Apprey, C.; Frimpomaa Agyapong, N.A. Nutritional knowledge and practices of mothers/caregivers and its impact on the nutritional status of children 6-59 months in Sefwi Wiawso Municipality, Western-North Region, Ghana. *Heliyon* **2022**, *8*, e12330. [CrossRef]
95. Debela, B.L.; Demmler, K.M.; Rischke, R.; Qaim, M. Maternal nutrition knowledge and child nutritional outcomes in urban Kenya. *Appetite* **2017**, *116*, 518–526. [CrossRef]
96. Gyampoh, S.; Otoo, G.E.; Aryeetey, R.N. Child feeding knowledge and practices among women participating in growth monitoring and promotion in Accra, Ghana. *BMC Pregnancy Childbirth* **2014**, *14*, 180. [CrossRef]
97. Ongosi, A.N. Nutrient Intake and Nutrition Knowledge of Lactating Women (0–6) Months Postpartum) in a Low Socio-Economic Area in Nairobi, Kenya. Ph.D. Thesis, University of Pretoria, Pretoria, South Africa, 2010.
98. Negash, C.; Whiting, S.J.; Henry, C.J.; Belachew, T.; Hailemariam, T.G. Association between Maternal and Child Nutritional Status in Hula, Rural Southern Ethiopia: A Cross Sectional Study. *PLoS ONE* **2015**, *10*, e0142301. [CrossRef]
99. Saaka, M. Relationship between mothers' nutritional knowledge in childcare practices and the growth of children living in impoverished rural communities. *J. Health Popul. Nutr.* **2014**, *32*, 237–248. Available online: https://www.ncbi.nlm.nih.gov/pmc/articles/PMC4216960/ (accessed on 29 June 2023).
100. Rohit, A.; Kirkham, R.; McCarthy, L.; Puruntatameri, V.; Maple-Brown, L.; Brimblecombe, J. Exploring differences in perceptions of child feeding practices between parents and health care professionals: A qualitative study. *BMC Public Health* **2021**, *21*, 1449. [CrossRef]
101. Modjadji, P.; Seabela, E.S.; Ntuli, B.; Madiba, S. Beliefs and Norms Influencing Initiation and Sustenance of Exclusive Breastfeeding: Experiences of Mothers in Primary Health Care Facilities in Ermelo, South Africa. *Int. J. Environ. Res. Public Health* **2023**, *20*, 1513. [CrossRef]
102. Madiba, S.; Modjadji, P.; Ntuli, B. Breastfeeding at Night Is Awesome; Mothers' Intentions of Continuation of Breastfeeding Extreme and Very Preterm Babies upon Discharge from a Kangaroo Mother Care Unit of a Tertiary Hospital in South Africa. *Healthcare* **2023**, *11*, 1048. [CrossRef]
103. Bhattacharjee, N.V.; Schaeffer, L.E.; Marczak, L.B.; Ross, J.M.; Swartz, S.J.; Albright, J.; Gardner, W.M.; Shields, C.; Sligar, A.; Schipp, M.F. Mapping exclusive breastfeeding in Africa between 2000 and 2017. *Nat. Med.* **2019**, *25*, 1205–1212. [CrossRef] [PubMed]
104. Cuartas, J.; Jeong, J.; Rey-Guerra, C.; McCoy, D.C.; Yoshikawa, H. Maternal, paternal, and other caregivers' stimulation in low-and- middle-income countries. *PLoS ONE* **2020**, *15*, e0236107. [CrossRef]
105. Kristo, A.S.; Sikalidis, A.K.; Uzun, A. Traditional Societal Practices Can Avert Poor Dietary Habits and Reduce Obesity Risk in Preschool Children of Mothers with Low Socioeconomic Status and Unemployment. *Behav. Sci.* **2021**, *11*, 42. [CrossRef]
106. Martorell, R.; Zongrone, A. Intergenerational influences on child growth and undernutrition. *Paediatr. Perinat. Epidemiol.* **2012**, *26* (Suppl. S1), 302–314. [CrossRef] [PubMed]
107. Doherty, T.; Sanders, D.; Jackson, D.; Swanevelder, S.; Lombard, C.; Zembe, W.; Chopra, M.; Goga, A.; Colvin, M.; Fadnes, L.T.; et al. Early cessation of breastfeeding amongst women in South Africa: An area needing urgent attention to improve child health. *BMC Pediatr.* **2012**, *12*, 105. [CrossRef] [PubMed]
108. Jama, N.A.; Wilford, A.; Masango, Z.; Haskins, L.; Coutsoudis, A.; Spies, L.; Horwood, C. Enablers and barriers to success among mothers planning to exclusively breastfeed for six months: A qualitative prospective cohort study in KwaZulu-Natal, South Africa. *Int. Breastfeed. J.* **2017**, *12*, 1–13. [CrossRef]
109. Madiba, S.; Langa, J. Cultural practices interfere with adherence to exclusive infant feeding: A qualitative study among HIV positive post natal women in Hammanskraal, South Africa: Child nutrition and feeding practices. *Afr. J. Phys. Health Educ. Recreat. Danc.* **2014**, *20*, 264–278. Available online: https://hdl.handle.net/10520/EJC162272 (accessed on 29 June 2023).
110. Mphego, Z.; Madiba, S.; Ntuli, B. The influence of the family on adherence to exclusive breastfeeding: Experiences of women living in extended family households in poorly resourced communities of Mpumalanga Province, South Africa: Child nutrition and feeding practices. *Afr. J. Phys. Health Educ. Recreat. Danc.* **2014**, *20*, 279–290. Available online: https://hdl.handle.net/10520/EJC162271 (accessed on 29 June 2023).

111. Ngoma-Hazemba, A.; Ncama, B.; Ahmed, Y. Mothers' Perception of Infant Feeding Counselling in The Context of Prevention of Mother-to-Child Transmission of HIV in Lusaka, Zambia. *Med. J. Zamb.* **2015**, *42*, 58–67. Available online: https://www.ajol.info/index.php/mjz/article/view/130013 (accessed on 29 June 2023).
112. World Medical Association. World Medical Association Declaration of Helsinki: Ethical principles for medical research involving human subjects. *JAMA* **2013**, *310*, 2191–2194. [CrossRef]

children

MDPI

Article

Association of Stunting with Socio-Demographic Factors and Feeding Practices among Children under Two Years in Informal Settlements in Gauteng, South Africa

Zandile Kubeka [1] and Perpetua Modjadji [1,2,*]

[1] Department of Public Health, School of Health Care Sciences, Sefako Makgatho Health Sciences University, 1 Molotlegi Street, Ga-Rankuwa, Pretoria 0208, South Africa

[2] Non-Communicable Diseases Research Unit, South African Medical Research Council, Tygerberg, Cape Town 7505, South Africa

* Correspondence: perpetua.modjadji@mrc.ac.za

Abstract: Despite improvements in childhood undernutrition through integrated nutritional programs in South Africa, stunting among children remains persistent, and is attributed to poor socio-demographic status. This context has been implicated in disrupting mothers' decisions regarding effective infant feeding, ultimately meaning that children's nutritional demands remain unmet. In view of this, we conducted a cross-sectional study to determine the association between socio-demographic factors and infant and young child feeding (IYCF) practices and stunting among children under two years receiving primary health care in informal settlements in Gauteng, South Africa. A validated questionnaire was used to assess mothers' socio-demographic status and feeding practices using WHO core indicators. Stunting was defined as length-for age z-scores (LAZ) below −2 standard deviation, computed using WHO Anthro software version 3.2.2.1 using age, sex, and anthropometric measurements of children. Univariate and multivariate analyses were stratified by stunting to determine the relationship with socio-demographic, infant, and IYCF factors using STATA 17. The prevalence of stunting was 16% among surveyed children under two years (with a mean age of 8 ± 5 months) living in poor socio-demographic households. Poor feeding practices were characterized by delayed initiation of breastfeeding (58%), sub-optimal exclusive breastfeeding (29%), discontinued breastfeeding (44%), early introduction of solid foods (41%), and low dietary diversity (97%). Significant differences in terms of child's age, monthly household income, and ever being breastfed were observed (Chi square test and univariate analysis). After controlling for potential confounders, stunting was significantly associated with child's age [12–23 months: AOR = 0.35, 95% CI: 0.16–0.76], and monthly household income [ZAR 3000–ZAR 5000: AOR = 0.47, 95% CI: 0.26–0.86]. Despite the few aforementioned socio-demographic and IYCF factors associated with stunting, this study reiterates stunting as the commonest poor nutritional status indicator among children under two years, suggesting the presence of chronic undernutrition in these poverty-stricken informal settlements. A multisectoral approach to address stunting should be context-specific and incorporate tailor-made interventions to promote optimal infant-feeding practices. Conducting future nutrient assessments focusing on children is imperative.

Keywords: stunting; socio-demographic factors; feeding practices; children; informal settlements; South Africa

1. Introduction

Infant and young child feeding (IYCF) is a set of practices recommended by the World Health Organization (WHO) to promote optimal child growth and development [1]. These feeding practices entail the initiation of breastfeeding within one hour of birth, exclusive breastfeeding for the first six months, and the introduction of nutritionally adequate and safe complementary foods at six months, together with a continuation of breastfeeding for up to two years of age or beyond [2]. Considering their importance, appropriate IYCF practices have been reported to prevent almost 19% of all deaths of under-fives [3]; however, inappropriate IYCF remains a concern in developing countries [2]. It is worth noting that these feeding practices are known to be affected by diverse factors such as family-related circumstances [4], parental knowledge, beliefs, and practices [5], food preparation and preservation, hygiene practices [6], and HIV infection, which can conflict with mothers' choices regarding breastfeeding [7].

In South Africa, mothers' feeding practices are sub-optimal, which form one of the major public health problems, despite the country having excellent policies in place and a political commitment to improve infant health and nutrition [8,9]. Studies have reported early breastfeeding initiated within one hour of birth (67.3%), a deficit in exclusive breastfeeding (31.6%), and the introduction of early complementary foods (85%) and mixed feeding (90%) [10,11]. Inappropriate feeding practices, especially among infants from six months of age transitioning from a milk diet to complementary foods, predispose them to unfavorable health outcomes and poor physical growth and development [12–14], which continue to exist and vary by sex and setting [12,13,15–17].

In particular, stunting (i.e., linear growth retardation or low length-for-age) remains the most common manifestation of undernutrition affecting children under five years amidst concerns regarding inappropriate feeding practices [17]. Local studies have reported stunting as the most persistent indicator for poor nutritional status among children, estimated at a prevalence between 45.3% and 55% [12,16]. Occasionally, stunting among children is accompanied by underweight (i.e., low weight-for-age) and thinness (i.e., low body mass index-for-height) [17], as well as overweight/obesity (i.e., overnutrition) [12,13,15,18–20]. The complexity of stunting characteristics can be explained through the multi-level factors contributing to stunting, which include poor socioeconomic status, poor maternal health, and nutrition, inadequate IYCF, and micronutrient deficiencies [21]. Stunted children experience a reduced lean body mass and diminished intellectual functioning, a short adult stature and reduced earnings later in life [22,23], and an increased risk of becoming overweight/obese in adulthood and developing non-communicable diseases (NCDs), which might be exacerbated by weight gain after the age of two years [24–26]. Therefore, due to various influences, a deficit in growth status in early life predisposes children to a slow linear growth during middle childhood [27].

Most concerning is the mushrooming of informal settlements in South Africa due to rapid urbanization, which makes this population susceptible to poverty and resultant ill health [28–31]. Informal settlements come with a poor infrastructure characterized by poor sanitation and electricity issues that make food storage difficult, as well as limited health services and other services, posing special risks to children [32]. Despite the spread of informal settlements, several public health issues, such as feeding practices and poor nutritional status, are less studied in these areas. However, we acknowledge the research conducted on cardiometabolic risks and quality of life [30,31,33–35]. This study determined the association between socio-demographic factors and IYCF practices and stunting among children under two years attending primary health care (PHC) facilities in informal settlements in Gauteng, South Africa. This study highlights the gap in IYCF practices in relation to the recommendations of the WHO among a group of children living in informal settlements. The importance of promoting optimal feeding practices to improve nutritional status of children is self-evident, as well as the need for efforts to strengthen contextual programs that educate mothers on feeding practices.

2. Materials and Methods

2.1. Study Design and Conceptual Frameworks

A cross-sectional study was conducted to determine the association of socio-demographic factors and IYCF practices with stunting among children under two years attending PHC facilities in the informal settlements in Gauteng, South Africa. The study adopted the core indicators for assessing optimal IYCF [36], entailing early initiation of breastfeeding, exclusive breastfeeding under six months, continues breastfeeding at 1 year, introduction of solid, semi solids, or soft foods, dietary diversity, minimum meal frequency, minimum acceptable diet, and consumption of iron-rich or iron-fortified foods. Additionally, the WHO framework for stunting was integrated explaining the factors predisposing to stunting from household and family, feeding practices, and infection [37]. The study was conducted between 2017 and 2018.

2.2. Study Setting and Population

This study was conducted at Ekurhuleni District situated in Gauteng Province, South Africa. Ekurhuleni is one of the five districts of Gauteng Province, and one of the eight metropolitan municipalities of South Africa with a total population of 3,500,231 based on the Mid-Year Population Estimates 2016 Statistics South Africa [38,39]. Ekurhuleni has about 91 PHCs, and 11 of them are in Ekurhuleni North 1 sub-district [38]. The study population included infants and their mothers attending PHCs for immunization and growth monitoring in Ekurhuleni North 1 clinics situated in the informal settlements. Gauteng province is one of three out of the nine provinces in South Africa with the highest number of households living in informal settlements [40]. Informal settlements are defined as unplanned residential areas where housing, shelter, and services have been constructed on land that the occupants occupy illegally, according to the United Nations [41]. These unplanned residential areas in South Africa are accompanied by poor health and socioeconomic status [29].

The study included children aged under two years who were brought to the PHC facilities by their biological mothers. Children diagnosed with medical conditions that would have interfered with their feeding and growth status and brought to PHC facilities by non-biological mothers were excluded from the study, as well as mothers who were not able to exclusively breastfeed due to health issues, not willing to participate in the study, and below the age of 18 years.

2.3. Sample Size and Sampling Procedure

Using the Cochran formula in a validated sample size calculator (Raosoft Inc., Seattle, WA, USA), a minimum recommended sample for this study was 370, and increased to 444 through a 20% buffering, to cater for a non-response. The total population of children attending PHC facilities in Ekurhuleni North 1, according to DHIS 2015/2016, was estimated to be about 9951 in all 11 clinics in Ekurhuleni north 1 sub-district. We considered an estimated population of children receiving care at PHC facilities, 95% confidence level, and a margin of error of 5% to estimate a representative sample of children in this sub-district [38]. In all, 450 mothers were recruited while waiting in queues for health services in the PHC facilities. Mothers who agreed to participate in the study were requested to come to the arranged space within the health facility after consultation for further engagements. The purpose of the study was explained to the selected mothers, prior to data collection. Out of the recruited 450 mothers, 448 responded, making a 99% response rate. At data capturing, we discovered that 15 of the participants were caregivers, and not biological mothers; meanwhile, at data analysis, six questionnaires had missing data above 10%, as a result, twenty-one questionnaires were excluded. The final sample size of the study was 427 mother–child pairs. First, the three clinics were purposively selected out of the eleven clinics in Ekurhuleni North 1 sub-district, based on the highest population headcount attending PHCs for immunization. Second, within the selected clinics, mothers were selected based on their availability. Hence, convenience sampling, applicable in clinical research [42], was used to select mothers and their infants, following the difficulty in obtaining a random sampling due to long queues in the health facilities, impatience, and resistance. Resistance, low participation of informal settlers in research, and community have been documented due to poor service delivery [43]. A flow chart for recruitment and sampling processes is presented in Figure 1.

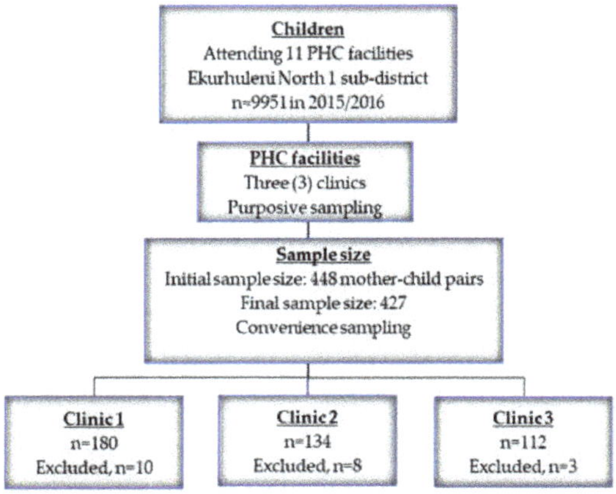

Figure 1. A flow chart on recruitment and sampling processes.

2.4. Data Collection and Tools

2.4.1. Feeding Practices

Data on infant feeding practices were collected using the WHO questionnaire, estimated from self-reported 24-h recall [44]. The eight core indicators for assessing optimal IYCF, which informed the data collected on infant feeding practices in the current study are [44]:

- "Early initiation of breastfeeding—a proportion of children born in the last 24 months who were put to the breast within one hour of birth) based on historic recall.
- Exclusive breastfeeding under 6 months—a proportion of infants 0–5 months who are fed exclusively with breastmilk, based on a 24-h recall/recall of the previous day.
- Continues breastfeeding at 1 year—a proportion of children 12–15 months who are fed breastmilk based on a recall of the previous day.
- Introduction of solid, semi solids or soft foods—a proportion of infants 6–8 months who received solid, semi solids or soft foods based on a recall of the previous day.
- Dietary diversity—a proportion of children 6–23 months who received foods from 4 or more food groups based on a recall of the previous day.
- Minimum meal frequency—minimum of 2 meals for infants 6–8 months and 3 meals for infants 9–12 months if breastfed; minimum of 4 meals aged 6–12 months if not breastfed.
- Minimum acceptable diet—infant $\geq$ 6 months meeting minimum requirements for dietary diversity and meal frequency.
- Consumption of iron-rich or iron-fortified foods—proportion of children 6–23 months of age who receive an iron-rich food or iron-fortified food that is specially designed for infants and young children, or that is fortified in the home during the previous day".

2.4.2. Socio-Demographic Variables

Socio-demographic data were collected using a structured questionnaire adapted from the literature [12,16]. Information on children's date of birth, gender, and illness history were collected in addition to mothers' household demographics and socioeconomic status through interviews conducted by the main researcher and trained research assistants. The questionnaire was validated through face and content validity, translation, and a pilot study [45]. The questionnaire was first prepared in English and then translated into isiZulu and Sesotho (i.e., a mixture of Sepedi and Setswana) to make it easy for the mothers to understand the questions. The question was further validated through content and face validity, and a pilot study [46]. Independent translators, who spoke isiZulu and Sesotho as their mother tongue and were conversant with English, completed the forward and backward translations of the questionnaire. An expert committee on public health nutrition studies approved the final version of the translated questionnaire. A pilot study was conducted to pre-test the questionnaire and determine its feasibility, among 30 mothers who did not participate in the main study, estimated to be a suitable number for piloting a concept. In addition to training the research assistants to administer the questionnaire and measure anthropometry, they were moderated while conducting preliminary interviews in local languages during piloting. Following a pilot study, only minimal changes to the study were implemented in terms of layout and style, and the results were not included in the main study.

2.4.3. Anthropometric Measurements and Nutritional Status Indicators of Children

Trained research assistants collected children's weights and heights according to WHO procedures [47]. Weight (W) was measured to the nearest 0.1 g using a Seca 354 baby electronic scale by Medicare hospital equipment in South Africa, manufactured in Germany. This measurement entailed the child's length lying down (recumbent) on the scale placed on a flat, stable surface such as a table. The recumbent length of each child was measured in a supine position to the nearest 0.1 cm using the Seca 210 measuring mat. Measurements were recorded on an anthropometry sheet and captured on the WHO Anthro software version 3.2.2.1 to computed Z-scores based on the 2006 WHO growth standards, expressed as standard deviation units from the median value for the WHO growth reference groups taking age and sex into consideration. From the length measured above from children, stunting was computed as length-for-age z-score (LAZ) below −2 standard deviation (SD). From the weight measured above from children, underweight was computed as weight-for-age z-scores (WAZ) below −2SD, and from the generated body mass index (BMI) of children, thinness was computed as BMI-for-age z-scored (BAZ) below −2SD, and overweight and obesity as BAZ > 2SD and ≤ 3SD, and BAZ > 3SD, respectively [47,48].

2.5. *Data Analysis*

STATA version 17 (Stata Corp. 2015. Stata Statistical Software: Release 17, College Station, TX, USA) was used to compute descriptive and inferential statistics. Missing data were detected through complete case analysis. Tabulations of household and maternal socio-demographic variables were performed through frequencies (n) and percentages. Distributions of continuous variables, such as age, anthropometry, and z-scores were checked using the Shapiro–Francia test, and results presented as median (interquartile range (IQR)]. Data analysis was stratified by stunting as the outcome variable; meanwhile, the numerical data were categorized for comparison purposes, easier interpretation, results, and the display of descriptive statistics. Non-parametric tests, Mann–Whitney U and Kruskal–Wallis, were used to compare the medians between two groups and three groups by sex and age groups, respectively. A Chi-square test (and Fischer's exact test for cells with less than five cases) were used to compare non-stunted and stunted children by mothers' and children's characteristics. Univariate [crude odds ratio (COR); p-value ≤ 0.25] and multivariate analyses [adjusted odd ratio (AOR); p-value < 0.05] were used to determine the associations of stunting and selected covariate.

3. Results

3.1. *The Nutritional Status of Children*

The study consisted of 427 mother–child pairs. Out of 427 children, 210 (49%) were boys and 217 (51%) were girls. Children were further divided into three age groups; younger (0–5 months; n = 176 (41%)), middle aged (6–11 months; n = 145 (34%)), and older (12–23 months; n = 106 (25%)) children with a mean age of 8 ± 5 months. In Figure 2, nutritional status indicators (i.e., stunting, underweight, and thinness) among children are plotted by sex and age groups. No significant differences were observed for median LAZ, WAZ, and BAZ by sex (using the Mann–Whitney test) and age (using the Kruskal–Wallis test).

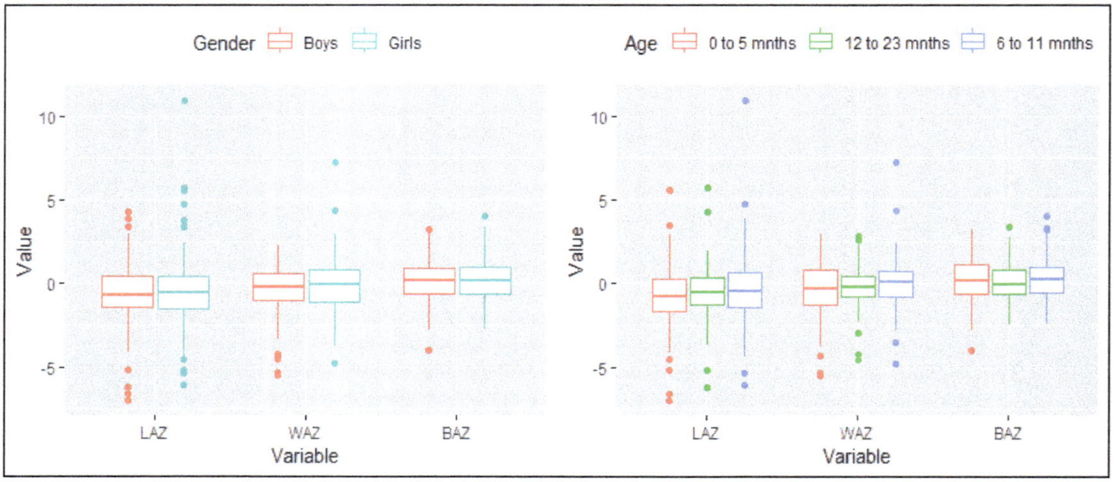

Figure 2. Stunting, underweight, and thinness among children by sex and age groups.

3.2. Prevalence of Stunting, Underweight and Thinness

Figure 3 describes and compares the prevalence of stunting, underweight, and thinness of children by age and sex groups using Chi-square test and Fischer's exact for cells with less than five cases. LAZ yielded the prevalence of children with normal z-scores, stunted, and tall, while WAZ yielded the prevalence of children with normal z-scores, underweight, and growth problems, and BAZ yielded the prevalence of children with normal z-scores, thinness, overweight risk, overweight, and obesity. During analysis, we considered normal children, and those with stunting, underweight, and thinness. In all, 16% of children were stunted, 10% were underweight, and 3% were thin. Prevalence of stunting (p = 0.017) and underweight (p = 0.020) were significantly different by age and by sex.

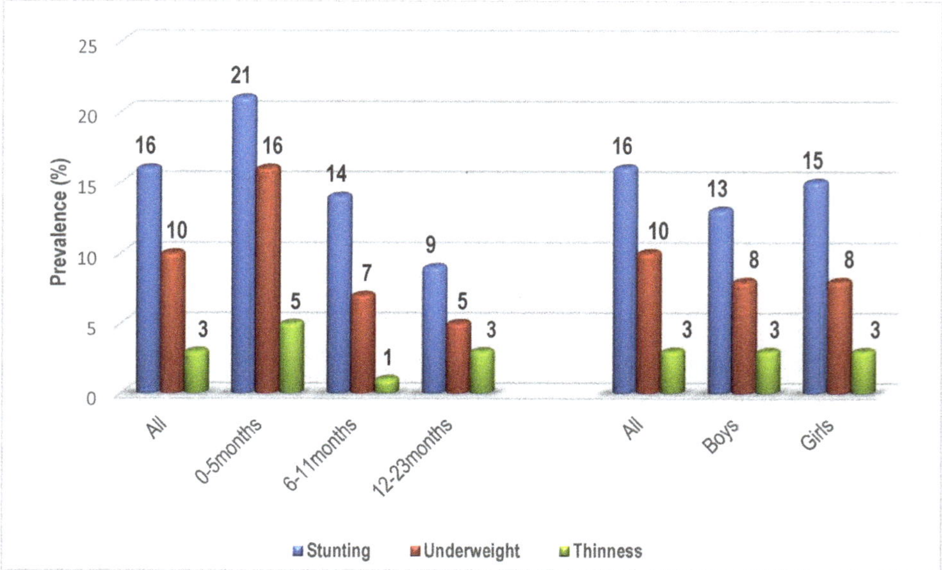

Figure 3. The prevalence of nutritional status indicators of children by age and sex groups.

3.3. Stunted/Non-Stunted Children and Characteristics of Mothers

Mothers were divided into two age groups; younger mothers aged below 30 years (n = 260 (61%)) and older mothers aged 30 years and above (n = 167 (39%)), with mean age of 29 ± 6 years ranging between 18 and 45 years. Most mothers were single (59%), had secondary education (70%), were unemployed (77%), and living in larger households (38%). Household income was categorized into three groups, which were low (<ZAR 3000), middle (ZAR 3000–ZAR 5000), and high (>ZAR 5000) income, and the low (40%) and middle (47%) income were common. We roughly considered that the median wage in South Africa is ≈ZAR 3000 per month, which is considered low [49]. LAZ (n = 427) yielded the prevalence of children with normal z-scores, stunted, and tall, and during analysis, we considered normal children (n = 345) and stunted (n = 65), and excluded tall children (n = 17). Stunted and non-stunted children were compared by the demographics of their mothers using Chi-square test and Fischer's exact test for cells with less than five cases. Significant difference of stunted children was observed by monthly household income (p = 0.024) and over half of stunted children (51%) lived in households with a low income compared to non-stunted children (39%) (Table 1).

Table 1. Comparison of stunted/non-stunted children by mothers' characteristics.

Variables	All (n = 427) n (%)	Non-Stunted (n = 345) n (%)	Stunted (n = 65) n (%)	p-Value
Mothers' age (years)				
<30	260 (61)	211 (61)	42 (65)	0.559
≥30	167 (39)	134 (39)	23 (35)	
Marital status				
Cohabiting	159 (37)	129 (37)	27 (42)	0.806
Single	252 (59)	203 (59)	36 (55)	
Ever married	16 (3)	13 (4)	2 (3)	
Education level				
No education/Primary	41 (10)	30 (9)	7 (11)	0 886
Secondary	301 (70)	245 (71)	45 (69)	
Tertiary	85 (20)	70 (20)	13 (20)	
Employment status				
Unemployed	330 (77)	265 (77)	54 (83)	0.265
Employed	97 (23)	80 (23)	11 (17)	
Household size				
1–4	264 (62)	211 (61)	41 (63)	0.771
≥5	163 (38)	134 (23)	24 (37)	
Household income/month				
Low	170 (40)	134 (39)	33 (51)	0.024 *
Middle	201 (47)	169 (49)	20 (31)	
High	12 (3)	42 (12)	12 (18)	
Type of house				
Brick	6 (1)	4 (1)	2 (3)	0.384
Prefabricated	8 (2)	8 (2)	0	
Corrugated/zinc	114 (27)	93 (27)	16 (25)	
Mud/wood	299 (70)	240 (70)	47 (72)	
Water access				
No	19 (4)	15 (4)	3 (5)	0.923
Yes	421 (96)	330 (96)	62 (95)	
Type of toilet				
Non-flush	85 (19)	66 (19)	12 (18)	0.900
Flush	342 (80)	279 (81)	53 (82)	

* Stands for significant difference, n stands for number of participants, and % stands for percentage.

3.4. Stunted/Non-Stunted Children and Their Characteristics

LAZ (n = 427) yielded the prevalence of children with normal z-scores, stunted, and tall, and during analysis using Chi-square test and Fischer's exact test for cells with less than five cases, we considered normal children (n = 345) and stunted (n = 65), and excluded tall children (n = 17). Most children (n = 415 (98%)) were born in health facilities, and only few (n = 47 (11%)) were attending day-care. Significant differences of stunted children were observed for child's age (p = 0.017) and being ever breastfed (p = 0.038) compared to the non-stunted children. Most stunted children were aged below 5 months (55%) and between 6 and 11 months (39%), while few stunted children (8%) were ever breastfed compared to the non-stunted (8%) (Table 2). Further results showed no significant differences were observed for IYCF practices between stunted and non-stunted children, but results showed that more of the non-stunted children compared to stunted children were initiated to breastfeeding within one hour of birth in this study. Feeding practices of children who were characterized by initiation of breastfeeding within one hour was observed in one-third of children (31%), exclusive breastfeeding (29%), early introduction of solid foods (41%), and low dietary diversity in almost all children.

Table 2. Comparison of stunted/non-stunted children by their characteristics.

Variables	All (n = 427) n (%)	Non-Stunted (n = 345) n (%)	Stunted (n = 65) n (%)	p-Value
Child's sex				
Boys	210 (49)	174 (50)	30 (46)	0.527
Girls	217 (51)	171 (50)	35 (54)	
Child's age (months)				
0–5	176 (41)	132 (38)	36 (55)	0.017 *
6–11	145 (29)	118 (34)	20 (31)	
12–23	106 (25)	95 (28)	9 (14)	
Child's place of birth				
Health facility	415 (98)	334 (97)	65 (100)	0.144
Non-health facility	12 (2)	11 (3)	0	
Child attending day-care				
No	380 (89)	306 (89)	60 (92)	0.388
Yes	47 (11)	39 (11)	5 (20	
Ever breastfed				
No	32 (7)	28 (8)	1 (2)	0.038 *
Yes	395 (93)	317 (92)	64 (98)	

* Stands for significant level at <0.05, p stands for probability level, n = number of participants and % stands for percentage.

3.5. Association of Stunting with Covariates

Table 3 shows the association of stunting and selected socio-demographic factors and IYCF practices variables. Univariate logistic regression (COR) showed associations of stunting with child's age, household income, initiation of breastfeeding, and being ever breastfed ($p \leq 0.25$). After controlling for potential confounders (socio-demographic, anthropometric, and feeding practice variables), the final hierarchical logistic regression showed significant associations of stunting with child' age [12–23 months: AOR = 0.35, 95% CI: 0.16–0.76], and monthly household income [middle-income: AOR = 0.47, 95% CI: 0.26–0.86]. No significant associations were observed between stunting and feeding practice variables in the multivariable model.

Table 3. Association of stunting and covariates.

Stunting	COR (95% CI)	*p*-Value	AOR (95% CI)	*p*-Value
Age (months)				
0–5	1		1	
6–11	0.62 (0.34–1.13)	0.120	0.62 (0.34–1.13)	0.120
12–23	0.35 (0.16–0.76)	0.008 *	0.35 (0.16–0.76)	0.008 *
Household income (monthly)				
<ZAR 3000	1		1	
ZAR 3000–R5000	0.48 (0.26–0.88)	0.017 *	0.47 (0.26–0.86)	0.014 *
>ZAR 5000	1.16 (0.55–2.45)	0.696	1.09 (0.51–2.33)	0.821

* Stands for significant level at <0.05, *p* stands for probability level, COR stands for crude odds ratio, AOR stands for adjusted odds ratio, CI stands for confidence interval, and % stands for percentage.

4. Discussion

Investigating stunting among children who may be at risk of compromised physical growth and development can never be over-emphasized as one of the public health concerns in South Africa due to its early and long-term consequences and implications. The first 1000 days of life from conception to two years of age are risky for growth and development among children because of the associative link with feeding practices that may either have positive or negative long-lasting consequences [50]. Therefore, in this study, we determined the association of socio-demographic factors and IYCF practices with stunting among children under two years attending PHC facilities in informal settlements in Gauteng, South Africa.

Prevalence of stunting (16%) among children was high in the district, despite nutritional programs put in place to improve the nutritional status of South African population groups [13,15,19,20]. This prevalence of stunting was either lower or higher in comparison with the prevalence reported in other local studies by settings and regions [12,13,16,18,20]. For instance, different levels from low, moderate to high prevalence of stunting among children have been recorded in Limpopo (3.7–45.3%) [12,18], Mpumalanga (51%) [20], North-West (29%) [13] and Gauteng (55%) [16] provinces. The same public health challenge of persistent moderate to high prevalence of stunting has been observed in African countries like Ethiopia (37.9%) [51], and Kenya (24%) [52]. The discrepancies in persistent stunting in this study can be attributed to sample size and/or setting (i.e., context), which obviously might have affected the outcome variable. The consequences of stunting such as poor cognitive and physical development and increased risk of chronic diseases in adulthood [53], make stunting one major issue of concern in South Africa that requires continuous attention, especially during the first two years of life [54].

Regarding feeding practices in this study, delayed initiation of breastfeeding (58%), sub-optimal exclusive breastfeeding (29%), and discontinuation of breastfeeding (44%) are consistent with local studies [14,55,56]. The breastfeeding initiation rate in this study is lower compared to rates reported in South Africa on a national level (88%) [57], in low-income areas (77%) [55], rural and peri-urban areas (88–100%) [14,58,59] and developing countries (95%) [60]. The importance of early breastfeeding initiation immediately after giving birth, establishing milk production, and increasing breastfeeding success has been acknowledged in South Africa [55,59]. It has been well documented that South Africa is behind on the global target of 50% exclusive breastfeeding, currently estimated at 32% [11]. Additionally, early introduction of solid foods (41%), and low dietary diversity (97%) in this study have been reported in other studies [12,13].

The differences between stunted and non-stunted children have been reported regarding growth and development [61], child's characteristics [62,63], as well as socio-demographic factors [37] and feeding practices [64]. In this study, we found associations of stunting with child's age and monthly household income, initiation of breastfeeding, and being ever breastfed (chi-square and univariate regression analysis tests), and child's age and monthly household income (multivariable regression analysis), similar to other reports [11,12,65–68]. For instance, children below 6 months of age (21%) were more stunted, followed by those aged between 6 and 11 (14%), and 12 and 23 months (9%). Consistent results have been reported in South Africa showing that stunting affected almost a third (28.5%) of 6-month-old infants, [56] and ranges from 11% to 16% among 6–12-month-old infants [69,70], with a similar trend in other African countries [71]. Previous studies in low-and-middle income countries have described that low height in stunted children is a main factor in growth and development delays [72,73]. Like other research linking stunting to breastfeeding [64], we also found that fewer children in the stunting group were ever breastfed compared with the not stunted group. Furthermore, more of non-stunted children compared to stunted children were initiated to breastfeeding within one hour of birth in this study. Literature documents the importance of breastfeeding to supply newborns with colostrum, a deep lemon-colored liquid secreted by the breasts for the first several postpartum days, and compared with mature milk, it (i.e., colostrum) contains more minerals and protein [74], which are necessary for forming the gut microbiota and immune system [61].

Lastly, on the contrary, inconsistencies have been reported regarding the associations of stunting with socio-demographic status, whereby some studies have reported no significant differences between the non-stunted and stunted children [75] while others have indicated clear significances [61,76,77]. But socio-economic factors, mostly including income, have consistently been implicated in the prevalence of stunting in developing countries [78]. We also found an association between household income and stunting in this study. This is not out of the ordinary, considering that stunting develops over a relatively long period and is attributable to household poverty [51]. Poverty in this study is explained by poor living conditions observed through poor socio-economic households' context, larger numbers of household members, and predominant low and middle household income. Poverty in South African households has been reported [17,79], especially in settings with poor infrastructure like rural and informal settlements [27,31]. Worth noting, is that South Africa has experienced urbanization, which gave rise to informal settlements, poverty, and resultant ill health [29,31]. To date, the country continues to experience poverty and poor socio-economic status, predisposing households to food insecurity through limited availability, access, and affordability [66,80]. Therefore, compromised household food security promotes low dietary diversity, also observed in this study, depriving children of the ability to meet their nutrient requirements for growth, in addition to suboptimal feeding practices implicated in the minimal research reported previously in informal settlements in South Africa [81]. Further reports in other African countries have associated stunting with low wealth index [82] and pointed out that involvement from a spouse in child caregiving and providing income could contribute to the better health of the children [83,84]. Therefore, the high rate of singlehood in this study substantiates lack of spousal support with harmful consequence towards household upkeeping regarding income and food security.

Limitations

Although the cross-sectional design used in this study rules out causality, inferences are well founded considering the reliability and validity maintained during the research process. Second, we used convenience sampling due the difficulty of obtaining a random sample, and this was mitigated by obtaining a large sample size for the study. Third, some possibly unmeasured confounders and residual confounding coupled with potential reporting and recalling bias and social desirability may have influenced the effect of feeding practices on stunting. Fourth, the limited associations between stunting and covariates may not necessarily be generalizable to every resource-constrained study but rather may be context-specific. We acknowledge not testing the reliability of the questionnaire statistically as one of the characteristics of a valid questionnaire; however, the questionnaire was adapted from credible studies and piloted before use to ensure its reliability. Finally, we did not measure the HIV status of mothers, which could have provided a better indication regarding breastfeeding, in addition to omitting mother's anthropometric measurements.

Nonetheless, our study has been able to determine the prevalence of stunting among children aged under two years in the PHC facilities in selected informal settlements in Gauteng, South Africa, and further associated stunting with the socio-demographic factors and IYCF practices.

5. Conclusions

This study reports a high prevalence of stunting among children under two years in selected informal settlements in Ekurhuleni District situated in Gauteng Province of South Africa. Stunting was associated with child's age, monthly household income, initiation of breastfeeding, and being ever breastfed. Further logistic regression analysis showed that children aged 12–23 months have lower odds of increased risk of stunting compared to children aged between 0 and 5 months. Children living in households with a middle income monthly are less likely to be stunted compared to children living in households with a low income. Despite the few aforementioned socio-demographic and IYCF factors associated with stunting, this study reiterates stunting as the most common poor nutritional status indicator among children under two years, suggesting the presence of chronic undernutrition in these poverty-stricken informal settlements. These findings highlight suspected growth and development delays among these children, who might further be predisposed to lower cognitive function than children who are not stunted because of their stunted brains, which affect their learning abilities and education performance in school as they grow, as well as lack of productivity as they age into adulthood. Therefore, possibilities are that the context of poverty observed in this study disrupts mothers from making decisions for effective infant feeding, ultimately condoning unmet child nutritional demands. The multisectoral approach to address stunting should be context-specific and incorporate tailor-made interventions to promote optimal infant feeding practices among mothers. Future research on nutrient assessments among children is imperative.

Author Contributions: Conceptualization, Z.K. and P.M.; methodology, Z.K. and P.M.; software, P.M.; validation, Z.K.; formal analysis, P.M. and Z.K.; investigation, Z.K.; resources, Z.K.; data curation, Z.K. and P.M., writing—original draft preparation, Z.K. and P.M.; writing—review and editing, P.M.; supervision, P.M.; project administration, Z.K. All authors have read and agreed to the published version of the manuscript.

Funding: This research received no external funding.

Institutional Review Board Statement: This study was conducted according to the guidelines laid down in the Declaration of Helsinki [85] and all procedures involving human subjects were approved by Sefako Makgatho Health Sciences University Research and Ethics Committee (SMUREC); (SMUREC/H/117/2017: PG) approved on the 1 June 2017. The study further received permission, from the Ekurhuleni Research Committee, in Gauteng province of South Africa (GP_2017RP8_732) on the 29 August 2017.

Informed Consent Statement: The nature of the study was explained to mothers who attended facilities with their children for childcare services prior to participating in the study. Informed consent was obtained from biological mother involved in the study.

Data Availability Statement: The dataset for the study group generated and analyzed during the current study is available from the corresponding author upon reasonable request due to ethical restrictions.

Acknowledgments: The authors wish to thank the mothers for participating in this study with their children. We also thank the Ekurhuleni Research Committee, in Gauteng Province, South Africa, and the managers of the health facilities for giving us the permission to conduct the study, and the research assistants for data collection.

Conflicts of Interest: The authors declare no conflict of interest.

References

1. World Bank. Repositioning Nutrition as Central to Development: A Strategy for Large-Scale Action. 2006. Available online: https://documents.worldbank.org/en/publication/documents-reports/documentdetail/185651468175733998/repositioning-nutrition-as-central-to-development-a-strategy-for-large-scale-action-overview (accessed on 12 July 2023).
2. Nguyen, P.H.; Menon, P.; Ruel, M.; Hajeebhoy, N. A situational review of infant and young child feeding practices and interventions in Viet Nam. *Asia Pac. J. Clin. Nutr.* **2011**, *20*, 359–374. [PubMed]
3. UNICEF Division of Communication. *Tracking Progress on Child and Maternal Nutrition: A Survival and Development Priority*; UNICEF: New York, NY, USA, 2009.
4. Talbert, A.; Jones, C.; Mataza, C.; Berkley, J.A.; Mwangome, M. Exclusive breastfeeding in first-time mothers in rural Kenya: A longitudinal observational study of feeding patterns in the first six months of life. *Int. Breastfeed. J.* **2020**, *15*, 17. [CrossRef] [PubMed]
5. WHO. *Infant and Young Child Feeding: Model Chapter for Textbooks for Medical Students and Allied Health Professionals*; World Health Organization: Geneva, Switzerland, 2009.
6. Saleh, F.; Ara, F.; Hoque, M.A.; Alam, M.S. Complementary feeding practices among mothers in selected slums of Dhaka city: A descriptive study. *J. Health Popul. Nutr.* **2014**, *32*, 89–96. [PubMed]
7. Brown, A.; Raynor, P.; Lee, M. Healthcare professionals' and mothers' perceptions of factors that influence decisions to breastfeed or formula feed infants: A comparative study. *J. Adv. Nurs.* **2011**, *67*, 1993–2003. [CrossRef] [PubMed]
8. Chakona, G. Social circumstances and cultural beliefs influence maternal nutrition, breastfeeding and child feeding practices in South Africa. *Nutr. J.* **2020**, *19*, 47. [CrossRef] [PubMed]
9. Du Plessis, L.M.; Kruger, H.S.; Sweet, L. Complementary feeding: A critical window of opportunity from six months onward. *S. Afr. J. Clin. Nutr.* **2013**, *26*, S129–S140.
10. Demographic and Health Surveys Program STAT Compiler: South Africa. Available online: https://wwwstatcompilercom/en/ (accessed on 28 February 2020).
11. Sayed, N.; Schönfeldt, H.C. A review of complementary feeding practices in South Africa. *S. Afr. J. Clin. Nutr.* **2020**, *33*, 36–43. [CrossRef]
12. Modjadji, P.; Mashishi, J. Persistent Malnutrition and Associated Factors among Children under Five Years Attending Primary Health Care Facilities in Limpopo Province, South Africa. *Int. J. Environ. Res. Public Health* **2020**, *17*, 7580. [CrossRef]
13. Modjadji, P.; Molokwane, D.; Ukegbu, P.O. Dietary Diversity and Nutritional Status of Preschool Children in North West Province, South Africa: A Cross Sectional Study. *Children* **2020**, *7*, 174. [CrossRef]
14. Mushaphi, L.; Mbhenyane, X.; Khoza, L.; Amey, A. Infant-feeding practices of mothers and the nutritional status of infants in the Vhembe District of Limpopo Province. *S. Afr. J. Clin. Nutr.* **2008**, *21*, 36–41. [CrossRef]
15. Modjadji, P.; Madiba, S. Childhood Undernutrition and Its Predictors in a Rural Health and Demographic Surveillance System Site in South Africa. *Int. J. Environ. Res. Public Health* **2019**, *16*, 3021. [CrossRef]
16. Modjadji, P.; Pitso, M. Maternal Tobacco and Alcohol Use in Relation to Child Malnutrition in Gauteng, South Africa: A Retrospective Analysis. *Children* **2021**, *8*, 133. [CrossRef]
17. NDoH; StatsSA; SAMRC; ICF. *South Africa Demographic and Health Survey 2016: Key Indicators*; Statistics South Africa: Pretoria, South Africa; Rockville, MD, USA, 2017.
18. Mamabolo, R.L.; Alberts, M.; Steyn, N.P.; Delemarre-van de Waal, H.A.; Levitt, N.S. Prevalence and determinants of stunting and overweight in 3-year-old black South African children residing in the Central Region of Limpopo Province, South Africa. *Public Health Nutr.* **2005**, *8*, 501–508. [CrossRef]
19. Modjadji, P.; Madiba, S. The double burden of malnutrition in a rural health and demographic surveillance system site in South Africa: A study of primary schoolchildren and their mothers. *BMC Public Health* **2019**, *19*, 1087. [CrossRef]
20. Modjadji, P.; Masilela, L.N.; Cele, L.; Mathibe, M.; Mphekgwana, P.M. Evidence of Concurrent Stunting and Obesity among Children under 2 Years from Socio-Economically Disadvantaged Backgrounds in the Era of the Integrated Nutrition Programme in South Africa. *Int. J. Environ. Res. Public Health* **2022**, *19*, 12501. [CrossRef]
21. Amaha, N.D.; Woldeamanuel, B.T. Maternal factors associated with moderate and severe stunting in Ethiopian children: Analysis of some environmental factors based on 2016 demographic health survey. *Nutr. J.* **2021**, *20*, 18. [CrossRef]
22. de Onis, M.; Branca, F. Childhood stunting: A global perspective. *Matern. Child Nutr.* **2016**, *12* (Suppl. 1), 12–26. [CrossRef]
23. Dewey, K.G.; Begum, K. Long-term consequences of stunting in early life. *Matern. Child Nutr.* **2011**, *7* (Suppl. 3), 5–18. [CrossRef]
24. Symington, E.A.; Gericke, G.J.; Nel, J.H.; Labadarios, D. The relationship between stunting and overweight among children from South Africa: Secondary analysis of the National Food Consumption Survey—Fortification Baseline I. *S. Afr. Med. J.* **2015**, *106*, 65–69. [CrossRef]
25. World Health Organization. *Global Nutrition Targets 2025: Stunting Policy Brief*; World Health Organization: Geneva, Switzerland, 2014.
26. Devkota, M.D.; Adhikari, R.K.; Upreti, S.R. Stunting in Nepal: Looking back, looking ahead. *Matern. Child Nutr.* **2016**, *12* (Suppl. 1), 257–259. [CrossRef]
27. Modjadji, P.; Madiba, S. The Multidimension of Malnutrition among School Children in a Rural Area, South Africa: A Mixed Methods Approach. *Nutrients* **2022**, *14*, 5015. [CrossRef] [PubMed]
28. Huchzermeyer, M.; Karam, A. *Informal Settlements: A Perpetual Challenge?* Juta and Company Ltd.: Cape Town, South Africa, 2006.

29. Sverdlik, A. Ill-health and poverty: A literature review on health in informal settlements. *Environ. Urban.* **2011**, *23*, 123–155. [CrossRef]
30. Williams, D.S.; Máñez Costa, M.; Sutherland, C.; Celliers, L.; Scheffran, J. Vulnerability of informal settlements in the context of rapid urbanization and climate change. *Environ. Urban.* **2019**, *31*, 157–176. [CrossRef]
31. Mokwena, K.; Modjadji, P. Pre-morbid cardiometabolic risks among South Africans living in informal settlements. *Cardiovasc. J. Afr.* **2023**, *34*, 23–29. [CrossRef] [PubMed]
32. Dangour, A.D.; Watson, L.; Cumming, O.; Boisson, S.; Che, Y.; Velleman, Y.; Cavill, S.; Allen, E.; Uauy, R. Interventions to improve water quality and supply, sanitation and hygiene practices, and their effects on the nutritional status of children. *Cochrane Database Syst. Rev* **2013**, *1*, Cd009382. [CrossRef]
33. Ballantyne, R.; Oelofse, C. Informal settlers' perceptions of environmental quality: A case study of the Mizamoyethu Community, South Africa. *Environmentalist* **1999**, *19*, 203–215. [CrossRef]
34. Richards, R.; O'Leary, B.; Mutsonziwa, K. Measuring quality of life in informal settlements in South Africa. *Soc. Indic. Res.* **2007**, *81*, 375–388. [CrossRef]
35. Weimann, A.; Oni, T. A Systematised Review of the Health Impact of Urban Informal Settlements and Implications for Upgrading Interventions in South Africa, a Rapidly Urbanising Middle-Income Country. *Int. J. Environ. Res. Public Health* **2019**, *16*, 3608. [CrossRef]
36. World Health Organization. *Indicators for Assessing Infant and Young Child Feeding Practices: Part 1: Definitions: Conclusions of a Consensus Meeting Held 6–8 November 2007 in Washington DC, USA*; World Health Organization: Geneva, Switzerland, 2008.
37. Stewart, C.P.; Iannotti, L.; Dewey, K.G.; Michaelsen, K.F.; Onyango, A.W. Contextualising complementary feeding in a broader framework for stunting prevention. *Matern. Child Nutr.* **2013**, *9* (Suppl. 2), 27–45. [CrossRef]
38. Massyn, N.; Peer, N.; English, R.; Paradarath, A.; Barron, P.; Day, C. *District Health Barometer 2017/18*; Health System Trust: Durban, South Africa, 2019; pp. 1–407.
39. South Africa Web DHIS. 2018. Available online: https://za.dhis.dhmis.org/dhis-web-pivot/index.html (accessed on 23 February 2021).
40. The Housing Development Agency (HDA). Gauteng: Informal Settlements Status. 2012. Available online: http://thehda.co.za/pdf/uploads/multimedia/HDA_Informal_settlements_status_Gauteng.pdf (accessed on 26 June 2022).
41. United Nations High Commission for Refugees. Global Strategy for Settlement and Shelter. 2014. Available online: http://www.unhcr.org/530f13aa9.pdf (accessed on 26 June 2022).
42. Elfil, M.; Negida, A. Sampling methods in Clinical Research; an Educational Review. *Emerg (Tehran)* **2017**, *5*, e52.
43. Ndlovu, N.S.M. Exploring the Planning Challenges of Service Delivery in Local Municipalities: The Case of the Midvaal Local Municipality. Ph.D. Thesis, University of the Witwatersrand, Johannesburg, South Africa, 2016.
44. WHO Multicentre Growth Reference Study Group. Complementary feeding in the WHO multicentre growth reference study. *Acta Paediatr.* **2006**, *450*, 27–37.
45. Perneger, T.V.; Courvoisier, D.S.; Hudelson, P.M.; Gayet-Ageron, A. Sample size for pre-tests of questionnaires. *Qual. Life Res.* **2015**, *24*, 147–151. [CrossRef]
46. Tsang, S.; Royse, C.F.; Terkawi, A.S. Guidelines for developing, translating, and validating a questionnaire in perioperative and pain medicine. *Saudi J. Anaesth.* **2017**, *11*, S80. [CrossRef]
47. World Health Organization. *Guideline: Assessing and Managing Children at Primary Health-Care Facilities to Prevent Overweight and Obesity in the Context of the Double Burden of Malnutrition*; World Health Organization: Geneva, Switzerland, 2017.
48. World Health Organization. *WHO Anthro for Personal Computers, Version 3.2. 2, 2011: Software for Assessing Growth and Development of the World's Children*; World Health Organization: Geneva, Switzerland, 2010.
49. Petermaritzburg Economic Justiceand Dignity Research Report. 2020. Available online: https://pmbejd.org.za/wp-content/uploads/2020/06/PMBEJD-Research-Report-23062020.pdf (accessed on 9 July 2023).
50. Schwarzenberg, S.J.; Georgieff, M.K. Advocacy for Improving Nutrition in the First 1000 Days to Support Childhood Development and Adult Health. *Pediatrics* **2018**, *141*, e20173716. [CrossRef]
51. Lisanu Mazengia, A.; Andargie Biks, G. Predictors of Stunting among School-Age Children in Northwestern Ethiopia. *J. Nutr. Metab.* **2018**, *2018*, 7521751. [CrossRef]
52. Mwaniki, E.W.; Makokha, A.N. Nutrition status and associated factors among children in public primary schools in Dagoretti, Nairobi, Kenya. *Afr. Health Sci.* **2013**, *13*, 39–46. [CrossRef]
53. Black, R.E.; Victora, C.G.; Walker, S.P.; Bhutta, Z.A.; Christian, P.; de Onis, M.; Ezzati, M.; Grantham-McGregor, S.; Katz, J.; Martorell, R.; et al. Maternal and child undernutrition and overweight in low-income and middle-income countries. *Lancet* **2013**, *382*, 427–451. [CrossRef]
54. Said-Mohamed, R.; Micklesfield, L.K.; Pettifor, J.M.; Norris, S.A. Has the prevalence of stunting in South African children changed in 40 years? A systematic review. *BMC Public Health* **2015**, *15*, 534. [CrossRef]
55. Goosen, C.; McLachlan, M.H.; Schübl, C. Infant feeding practices during the first 6 months of life in a low-income area of the Western Cape Province. *S. Afr. J. Child Health* **2014**, *8*, 50–54. [CrossRef]
56. Matsungo, T.M.; Kruger, H.S.; Faber, M.; Rothman, M.; Smuts, C.M. The prevalence and factors associated with stunting among infants aged 6 months in a peri-urban South African community. *Public Health Nutr.* **2017**, *20*, 3209–3218. [CrossRef]

57. Shisana, O.; Rehle, T.; Simbayi, L.; Zuma, K.; Jooste, S.; Zungu, N.; Labadarios, D.; Onoya, D.; Wabiri, N. *South African National HIV Prevalence, Incidence and Behaviour Survey*; Human Sciences Research Council: Pretoria, South Africa, 2014.
58. Mamabolo, R.L.; Alberts, M.; Mbenyane, G.X.; Steyn, N.P.; Nthangeni, N.G.; Delemarre-Van De Waal, H.A.; Levitt, N.S. Feeding practices and growth of infants from birth to 12 months in the central region of the Limpopo Province of South Africa. *Nutrition* **2004**, *20*, 327–333. [CrossRef] [PubMed]
59. Sibeko, L.; Dhansay, M.A.; Charlton, K.E.; Johns, T.; Gray-Donald, K. Beliefs, attitudes, and practices of breastfeeding mothers from a periurban community in South Africa. *J. Hum. Lact.* **2005**, *21*, 31–38. [CrossRef]
60. Black, R.E.; Allen, L.H.; Bhutta, Z.A.; Caulfield, L.E.; de Onis, M.; Ezzati, M.; Mathers, C.; Rivera, J. Maternal and child undernutrition: Global and regional exposures and health consequences. *Lancet* **2008**, *371*, 243–260. [CrossRef]
61. Mustakim, M.R.D.; Irwanto; Irawan, R.; Irmawati, M.; Setyoboedi, B. Impact of Stunting on Development of Children between 1–3 Years of Age. *Ethiop. J. Health Sci.* **2022**, *32*, 569–578. [PubMed]
62. Berhe, K.; Seid, O.; Gebremariam, Y.; Berhe, A.; Etsay, N. Risk factors of stunting (chronic undernutrition) of children aged 6 to 24 months in Mekelle City, Tigray Region, North Ethiopia: An unmatched case-control study. *PLoS ONE* **2019**, *14*, e0217736. [CrossRef] [PubMed]
63. Woldehanna, T.; Behrman, J.R.; Araya, M.W. The effect of early childhood stunting on children's cognitive achievements: Evidence from young lives Ethiopia. *Ethiop. J. Health Dev. = Ya'Ityopya Tena Lemat Mashet* **2017**, *31*, 75–84.
64. Kuchenbecker, J.; Jordan, I.; Reinbott, A.; Herrmann, J.; Jeremias, T.; Kennedy, G.; Muehlhoff, E.; Mtimuni, B.; Krawinkel, M.B. Exclusive breastfeeding and its effect on growth of Malawian infants: Results from a cross-sectional study. *Paediatr. Int. Child Health* **2015**, *35*, 14–23. [CrossRef]
65. Modjadji, P.; Seabela, E.S.; Ntuli, B.; Madiba, S. Beliefs and Norms Influencing Initiation and Sustenance of Exclusive Breastfeeding: Experiences of Mothers in Primary Health Care Facilities in Ermelo, South Africa. *Int. J. Environ. Res. Public Health* **2023**, *20*, 1513. [CrossRef]
66. Modjadji, P. Engaging Mothers on the Growth of School-Age Children in a Rural South African Health and Demographic Site: A Qualitative Insight. *Healthcare* **2021**, *9*, 225. [CrossRef]
67. Zere, E.; McIntyre, D. Inequities in under-five child malnutrition in South Africa. *Int. J. Equity Health* **2003**, *2*, 7. [CrossRef]
68. Slemming, W.; Kagura, J.; Saloojee, H.; Richter, L.M. Early life risk exposure and stunting in urban South African 2-year old children. *J. Dev. Orig. Health Dis.* **2017**, *8*, 301–310. [CrossRef]
69. Faber, M.; Benade, A.J.S. Breastfeeding, complementary feeding and nutritional status of 6-12-month-old infants in rural KwaZulu-Natal. *S. Afr. J. Clin. Nutr.* **2007**, *20*, 16–24. [CrossRef]
70. Smuts, C.; Faber, M.; Schoeman, S.; Laubscher, J.; Oelofse, A.; Benadé, A.; Dhansay, M. Socio-demographic profiles and anthropometric status of 0- to 71-month-old children and their caregivers in rural districts of the Eastern Cape and KwaZulu-Natal provinces of South Africa. *S. Afr. J. Clin. Nutr.* **2008**, *21*, 117–124. [CrossRef]
71. Nshimyiryo, A.; Hedt-Gauthier, B.; Mutaganzwa, C.; Kirk, C.M.; Beck, K.; Ndayisaba, A.; Mubiligi, J.; Kateera, F.; El-Khatib, Z. Risk factors for stunting among children under five years: A cross-sectional population-based study in Rwanda using the 2015 Demographic and Health Survey. *BMC Public Health* **2019**, *19*, 175. [CrossRef]
72. Muhoozi, G.K.; Atukunda, P.; Mwadime, R.; Iversen, P.O.; Westerberg, A.C. Nutritional and developmental status among 6- to 8-month-old children in southwestern Uganda: A cross-sectional study. *Food Nutr. Res.* **2016**, *60*, 30270. [CrossRef]
73. Sudfeld, C.R.; McCoy, D.C.; Danaei, G.; Fink, G.; Ezzati, M.; Andrews, K.G.; Fawzi, W.W. Linear growth and child development in low- and middle-income countries: A meta-analysis. *Pediatrics* **2015**, *135*, e1266–e1275. [CrossRef]
74. Lockshin, M.D.; Sammaritano, L.R.; Schwartzman, S. 22—LUPUS PREGNANCY. In *Systemic Lupus Erythematosus*, 4th ed.; Lahita, R.G., Ed.; Academic Press: San Diego, CA, USA, 2004; pp. 659–695. [CrossRef]
75. Walker, S.P.; Chang, S.M.; Powell, C.A. The association between early childhood stunting and weight status in late adolescence. *Int. J. Obes.* **2007**, *31*, 347–352. [CrossRef]
76. Cabrera, N.J.; Volling, B.L.; Barr, R. Fathers are parents, too! Widening the lens on parenting for children's development. *Child Dev. Perspect.* **2018**, *12*, 152–157. [CrossRef]
77. Schady, N. Parents' education, mothers' vocabulary, and cognitive development in early childhood: Longitudinal evidence from Ecuador. *Am. J. Public Health* **2011**, *101*, 2299–2307. [CrossRef]
78. Keino, S.; Plasqui, G.; Ettyang, G.; van den Borne, B. Determinants of stunting and overweight among young children and adolescents in sub-Saharan Africa. *Food Nutr. Bull.* **2014**, *35*, 167–178. [CrossRef]
79. Gumede, V. Poverty in South Africa. 2021. Available online: https://www.researchgate.net/publication/353946196_Poverty_in_South_Africa (accessed on 15 August 2022).
80. Masuku, M.; Selepe, M.; Ngcobo, N. The socio-economic status as a factor affecting food (In) security in rural areas, uThungulu district municipality, Kwa-Zulu Natal, South Africa. *J. Hum. Ecol.* **2017**, *58*, 57–66. [CrossRef]
81. Drimie, S.; Faber, M.; Vearey, J.; Nunez, L. Dietary diversity of formal and informal residents in Johannesburg, South Africa. *BMC Public Health* **2013**, *13*, 911. [CrossRef] [PubMed]
82. Bommer, C.; Vollmer, S.; Subramanian, S.V. How socioeconomic status moderates the stunting-age relationship in low-income and middle-income countries. *BMJ Glob. Health* **2019**, *4*, e001175. [CrossRef] [PubMed]

83. Habimana, J.d.D.; Uwase, A.; Korukire, N.; Jewett, S.; Umugwaneza, M.; Rugema, L.; Munyanshongore, C. Prevalence and Correlates of Stunting among Children Aged 6–23 Months from Poor Households in Rwanda. *Int. J. Environ. Res. Public Health* **2023**, *20*, 4068. [CrossRef]
84. Garcia, I.L.; Fernald, L.C.H.; Aboud, F.E.; Otieno, R.; Alu, E.; Luoto, J.E. Father involvement and early child development in a low-resource setting. *Soc. Sci. Med.* **2022**, *302*, 114933. [CrossRef]
85. WMA. Declaration of Helsinki. Ethical principles for medical research involving human subjects. *Jahrb. Für Wiss. Ethik* **2009**, *14*, 233–238. [CrossRef]

children

MDPI

Article

Investigation of the Seasonal Variation in Fat Patterning of Ellisras Rural Children and Adolescents, in the Limpopo Province, South Africa

Lusani Mulaudzi [1], Thandiwe N. Mkhatshwa [1], Mankopodi M. Makhubedu [1], Moloko Matshipi [1], Sogolo L. Lebelo [2] and Kotsedi D. Monyeki [1,*]

[1] Department of Physiology and Environmental Health, University of Limpopo, Sovenga 0727, South Africa; 201709714@keyaka.ul.ac.za (L.M.); thandiwemngunomhle@gmail.com (T.N.M.); mankopodi.makhubedu@ul.ac.za (M.M.M.); moloko.matshipi@ul.ac.za (M.M.)

[2] School of Agriculture and Life Sciences, University of South Africa, Roodepoort 1710, South Africa; lebelol@unisa.ac.za

* Correspondence: kotsedi.monyeki@ul.ac.za; Tel.: +27-152682953

Abstract: An increase in childhood obesity has become a global concern since childhood obesity often leads to adulthood obesity. This study aimed to investigate whether there is seasonal variation in fat patterning, and whether fat patterning is associated with seasons among the Ellisras population aged 5–15 years. A total of 1986 children and adolescents (1034 boys and 952 girls) aged 5–15 enrolled in this cross-sectional study. Skinfold measurements were obtained using standard procedures. Obesity prevalence was determined using frequency analysis. The correlation between obesity and two seasons was tested using multinomial regression analysis. The results showed that girls aged 11–15 years had significantly ($p \leq 0.03$) larger median fat variables (triceps, biceps, and suprailiac) in spring compared to autumn. The prevalence of obesity ranged from 3–11% with boys being more obese (4–11%) than girls (3–7%) aged 5–15. Regression analysis showed a negative significant ($p \leq 0.001$) correlation between autumn and obesity by the subscapular/(subscapular + triceps) (SST) ratio, both unadjusted −2.991 (95%CI −3.233:−2.803) and adjusted −2.897 (95% CI −4.331:−1.629). The findings of the study showed that there was seasonal variation in fat patterning among boys and girls in the Ellisras population and that fat patterning is associated with seasons.

Keywords: fat patterning; seasonal variation; rural children and adolescents; South Africa

1. Introduction

Fat patterning refers to the distribution of adipose tissue throughout the human body. Developed and developing countries are experiencing an increase in childhood obesity, which has emerged as a risk factor for non-communicable diseases [1]. Therefore, the World Health Organisation (WHO) emphasises the need to monitor obesity in different populations [2]. Globally, the prevalence of obese individuals among adolescents has significantly increased in the last 20 years [3]. The prevalence of obesity among African children was reported to be 8.5% and was estimated to be 12.7% in 2020 [4].

Childhood and adolescent obesity are strongly influenced by risk factors such as nutrition, physical activity, the environment, socio-environmental factors, biological factors, and socio-economic status [5]. Nutritional imbalance in early life and a fat-accumulation-related lifestyle are crucial in the development of obesity [6]. Skinfold measurement is frequently used to measure body fat in children. Skinfold ratios indicate the subjects with the highest risk of obesity [7]. Childhood obesity that continues into maturity increases the possibility of developing weight-related illnesses including hypertension later in life [7]. This makes tracking and controlling obesity in children a priority [8]. Perinatal life, infancy, early childhood, and adolescence have been reported as critical stages for the development of obesity [9].

Most rural communities in developing nations have limited financial means for food availability and rely on their agricultural practices for production. Rainfall is essential for a productive harvest, and agricultural activities are heavily reliant on it and other climatic factors, which have an impact on food production and availability. Changes in rainfall and climatic conditions then result in seasonal food shortages, whereby food production is not enough to cover the population [10]. Therefore, people change their diets and eating habits to adapt to food shortages because of seasonal variations [10]. Other than environmental factors, socioeconomic status may also affect food patterns and food intake. It is a fact that studies that analyse dietary intake in most instances ignore the possibility of seasonal variability.

There is plenty of information that suggests that seasons could affect health conditions and therefore should be strongly considered in the incidence of nutrient deficiency and various food profiles [11]. Seasonal variations that occur throughout the year can alter the body's composition and fat distribution patterns. A study of seasonal variation in fat patterning amongst adolescents in urban areas showed that adolescents gain more weight during summer compared to winter [12]. However, there is limited information regarding weight changes among different seasons in South African children. Therefore, this study aimed to investigate seasonal variation in fat patterning in the rural Ellisras population and to investigate whether there is an association between fat patterning and seasons.

2. Materials and Methods

2.1. The Sampling Procedure

This study took place in Ellisras (now known as Lephalale), Limpopo province, South Africa. Ellisras is about 70 km from the Botswana border [13]. Most residents in Ellisras rely mostly on income from the Iscor coal mine and the Medupi power station. Fewer people are educated in this area; therefore, some depend on farming and cattle-rearing as a source of income. The spring season in Lephalale is said to be hot and reasonably dry with a daytime maximum temperature being around the average of 31 degrees Celsius (°C) while the nighttime maximum temperature is around an average of 18 °C. The autumn season is said to be mild and very dry with a maximum average daytime temperature of 26 °C and a maximum average temperature of 9 °C [14]. However, due to climate change, these weather conditions may not give a true reflection of the weather conditions in Lephalale in the year 2000 when data collection for this study took place.

2.2. Data Collection

This study is a subset of the Ellisras Longitudinal Study (ELS) and that is where data from this study was sourced [15,16]. A total of 1986 participants (1034 boys and 952 girls) aged 5–15 enrolled in this cross-sectional study. Measurements were taken in two seasons autumn and spring (May 2000, November 2000). The Ellisras Longitudinal Study (ELS) initially followed a cluster sampling procedure documented elsewhere [15,16]. Briefly, the study commenced with a total of 22 schools (10 preschools and 12 primary schools) that were randomly selected from 68 schools in the Ellisras area. A total of 2225 children (550 preschool learners and 1675 primary school learners) enrolled in the ELS at baseline in 1996 [15,16]. The ELS is an ongoing study and the same population that enrolled in the ELS at baseline is followed up to date.

This study used data collected in 2000 which is one of the oldest datasets from the ELS. Since the ELS follows the same participants over time, the 2000 data will allow tracking of the ELS participants from childhood into adulthood. The participants in the year 2000 were between 5–15 years of age, and currently all ELS participants are in their adulthood. Furthermore, the 2000 data did not have a lot of missing values for all skinfolds (triceps, biceps, subscapular, and suprailiac) measures in most of the participants. Participants with missing values for skinfold measurements were excluded from the analysis, with children less than 5 years and participants above 15 years excluded as well. Participants who did not submit a signed consent form were excluded also. All participants who were present on the day of data collection formed part of the study.

2.2.1. Anthropometry

The International Society for the Advancement of Kinanthropometry (ISAK) standard procedure was followed for training and measurements [17]. All children and adolescents underwent skinfold measurements using a Slim Guide skinfold calliper (Harpenden, England). Skinfolds (triceps, biceps, subscapular, and suprailiac) were measured three times, and the results were rounded off to the nearest 0.1 mm. For all skinfold measures, the intra-tester and inter-tester technical error of measurements ranged from 0.2–6 mm (0.4–6.8%), respectively, to ensure the study's validity and accuracy during data collection.

Triceps

Triceps were measured over the triceps muscle, on the posterior mid-acromiale-radiale line on a relaxed arm, with the shoulder joint slightly rotated externally as shown in Figure 1 [17].

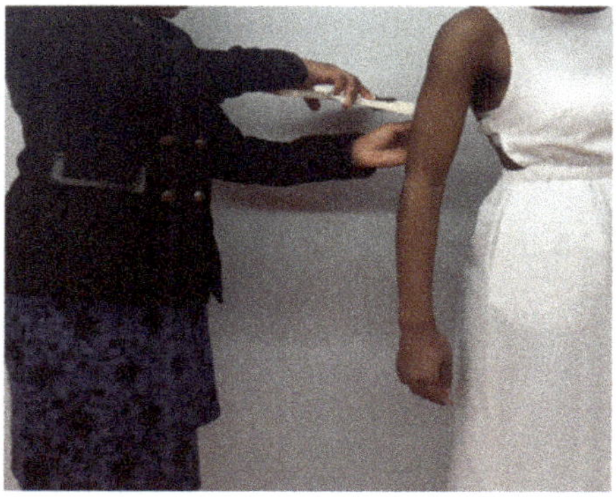

Figure 1. Triceps measurement.

Biceps

Biceps were measured over the triceps muscle, on the anterior mid-acromiale-radiale line of a relaxed arm, with the shoulder joint slightly rotated externally as shown in Figure 2 [17].

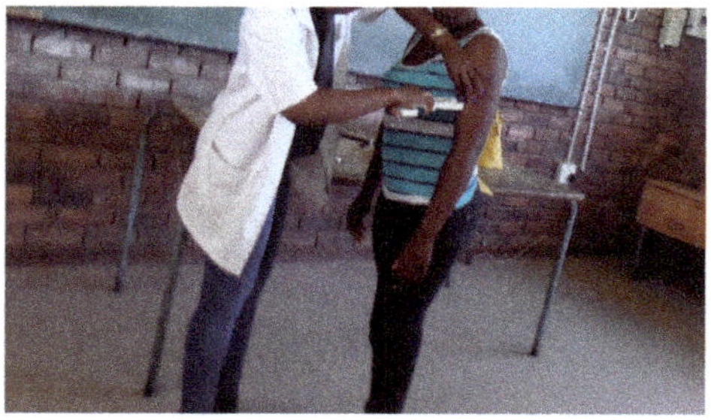

Figure 2. Biceps measurement.

Subscapular

The subscapular skinfold was measured by first determining the under- most tip of the scapula, by palpating the interior-most angle of the scapula. The skinfold was then raised and measured at a line running laterally and down the subscapular landmark as shown in Figure 3 [17].

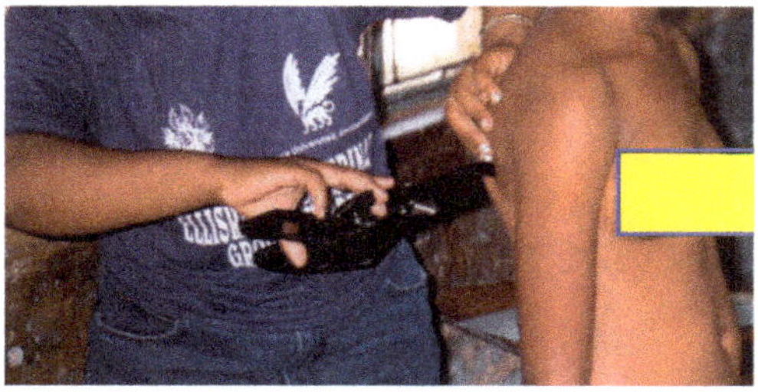

Figure 3. Subscapular measurement.

Suprailiac

The skinfold was measured on the ilio-axilla line superior to the iliocristale, with the arm abducted horizontally across the chest as shown in Figure 4 [17].

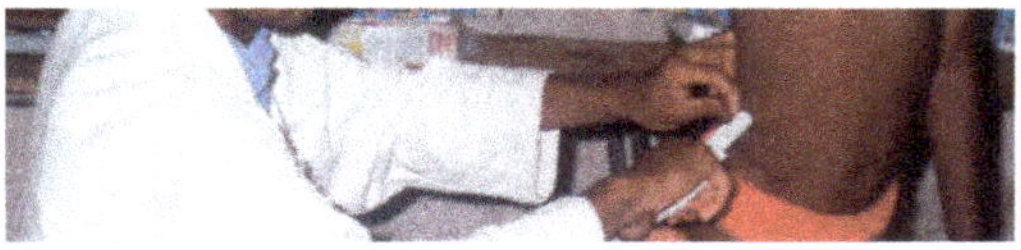

Figure 4. Suprailiac measurement.

2.3. Data Analysis

Data analysis was computed using IBM Statistical Package for the Social Sciences (Version 29) with a significance level set at $p \leq 0.05$. By comparing the ratio of trunk to limb body fat, central body fat distribution was identified. This was estimated utilising three equations [9].

$$\text{ST ratio} = \frac{\text{Subscapular}}{\text{triceps}}$$

$$\text{SST ratio} = \frac{\text{subscapular}}{\text{subscupular} + \text{triceps}}$$

$$\text{SSTB ratio} = \frac{\text{subscapular} + \text{suprailiac}}{\text{biceps} + \text{triceps}}$$

All subjects were classified as obese or normal using the 95th percentile cut-off points for the sum of four skinfolds (S4SK) for age and gender. The prevalence of obesity was determined using frequency analysis, and a nonparametric t-test was used for gender comparison. Participants were divided into age groups taking into consideration the critical stages for the development of obesity [18]. The association between obesity and seasons was investigated using the multinomial logistic regression analysis.

2.4. Ethical Clearance

The Turfloop Research Ethics Committee of the University of Limpopo granted ethical approval (clearance number: TREC/P/356/2017: PG) before the study was undertaken. Authorities and the principal of the school granted permission. The parents of the participants read and signed consent forms allowing participants to be part of the project.

3. Results

3.1. Characteristics of the Population

The results are shown as median and interquartile because the data were not normally distributed. Table 1 shows descriptive statistics of Ellisras rural children and adolescents measured in autumn and spring. Girls had significantly higher biceps measurements (6.75) than boys (4.0) at age 11–15 years ($p \leq 0.03$) in spring. Girls also had a significantly higher triceps measurement (9.0) than boys (6.75) at age 11–15 years ($p \leq 0.05$) in autumn.

Girls showed a significantly higher suprailiac measurement (6.75) than boys (4.25) at age 11–15 years ($p \leq 0.03$) in spring.

Table 1. Descriptive statistics of Ellisras rural children and adolescents measured in autumn and spring (May 2000, November 2000).

Variables	N	Boys Autumn 2000	N	Boys Spring 2000	N	Girls Autumn 2000	N	Girls Spring 2000
				Median (25:75 IQR)				
				Biceps				
5–10 years	413	4.0 ** (3.0:4.0)	335	4.0 ** (3.5:4.5)	378	5.5 * (4.5:6.5)	301	6.0 (4.5:7.0)
11–15 years	593	3.5 ** (3.0:4.0)	601	4.0 ** (3.5:4.5)	540	6.0 ** (5.0:7.0)	575	6.75 ** (5.5:8.0)
				Subscapular				
5–10 years	413	5.0 (5.0:6.0)	335	5.0 (5.0:6.0)	378	6.0 ** (5.0:7.0)	301	6.5 ** (6.0:7.5)
11–15 years	593	5.5 (5.0:6.25)	599	6.0 (5.0:6.5)	540	7.0 ** (6.0:8.0)	575	8.0 ** (6.75:9.0)
				Triceps				
5–10 years	413	6.5 (5.5:7.5)	335	6.5 (6.0:7.5)	378	7.88 (6.5:9.0)	301	8.0 (6.5:9.0)
11–15 years	593	6.75 ** (6.0:8.0)	599	7.0 ** (6.0:8.0)	540	9.0 * (7.5:11)	575	9.75 * (8.0:11.5)
				Suprailiac				
5–10 years	413	4.0 (3.0:4.0)	335	4.0 (3.5:4.5)	378	5.0 (4.0:6.0)	301	5.0 (4.0:6.0)
11–15 years	593	4.0 ** (3.5:5.0)	599	4.25 ** (4.0:5.0)	540	6.0 ** (5.0:8.0)	575	6.75 ** (5.5:8.0)

** Shows significance level where $p \leq 0.03$, * $p \leq 0.05$.

3.2. The Prevalence of Obesity Based on ST, SST, and SSTB Ratio

Table 2 displays the prevalence of obesity among Ellisras rural children and adolescents measured in autumn and spring. The prevalence of obesity ranged from 3–11% with boys being more obese (4–11%) than girls (3–7%) aged 5–15. The prevalence of obesity was mostly significantly higher in spring compared to autumn. Obesity by ST ratio among boys aged 11–15 years in 2000 was significantly ($p \leq 0.001$) higher in autumn (11%) compared to spring (8%). The prevalence of obesity by SSTB among girls 5–10 was significantly ($p \leq 0.001$) higher in spring (7%) compared to autumn (3%).

Table 2. Prevalence of fat patterning of Ellisras rural children and adolescents measured in autumn and spring (May 2000, November 2000).

Variables	N	Boys Autumn 2000	N	Boys Spring 2000	N	Girls Autumn 2000	N	Girls Spring 2000
		n (%)		n (%)		n (%)		n (%)
				Abdominal obesity by ST ratio				
5–10 years	413	33 (8)	335	24 (7)	378	17 (5) *	301	20 (7) *
11–15 years	593	63 (11) **	599	45 (8) **	540	16 (3) **	575	41 (7) **
				Abdominal obesity by SST ratio				
5–10 years	413	22 (5)	335	22 (7)	378	19 (5)	301	16 (5)
11–15 years	593	25 (4) **	599	43 (7) **	540	26 (5)	575	33 (6)
				Obesity by SSTB ratio				
5–10 years	413	23 (6)	335	19 (6)	378	13 (3) **	130	20 (7) **
11–15 years	593	29 (5)	599	33 (6)	540	30 (6)	575	26 (5)

* $p \leq 0.05$, ** Shows significance level where $p \leq 0.03$.

3.3. The Correlation of ST, SST, and SSTB Ratio with Season

Table 3 shows the multinomial regression in autumn (2000) and spring (2000). The regression analysis showed a significant negative association between seasons and obesity by ST, SST, and SSTB ratio ranging from −3.088 to −1.084. Obesity by ST ratio showed a significant ($p \leq 0.001$) negative correlation with spring when unadjusted −2.559 (95% CI −2.738:−2.382). There was a significant ($p \leq 0.001$) negative correlation between obesity by SST ratio and autumn both unadjusted −2.991 (95% CI −3.233:−2.803) and adjusted for age and gender −2.897 (95% CI −4.331:−1.629). Obesity by SSTB ratio showed a negative significant ($p \leq 0.001$) correlation with spring both unadjusted −2.860 (95%CI −3.087:−2.672) and adjusted for age and gender −2.506 (95% CI −3.885:−1.346).

Table 3. Multinomial logistic regression for the association between obesity and seasons (autumn and spring 2000).

	Autumn			Spring		
	Unadjusted					
	Beta	***p* Value**	**95% CI**	**Beta**	***p* Value**	**95% CI**
Abdominal obesity based on ST ratio	−2.633	<0.001 **	−2.822:−2.449	−2.559	<0.001 **	−2.738:−2.382
Abdominal obesity based on the SST ratio	−2.991	<0.001 **	−3.233:−2.803	−2.700	<0.001 **	−2.393:−2.518
Abdominal obesity based on the SSTB ratio	−2.958	<0.00 **	−3.191:−2.756	−2.860	<0.001 **	−3.087:−2.672
	Adjusted for age and gender					
Abdominal obesity based on ST ratio	−1.084	0.054	−2.302:−0.075	−2.556	<0.001 **	−3.903:−1.432
Abdominal obesity based on the SST ratio	−2.897	<0.001 **	−4.331:−1.629	−2.718	<0.001 **	−4.165:−1.501
Abdominal obesity based on the SSTB ratio	−3.088	<0.001 **	−4.573:−1.828	−2.506	<0.001 **	−3.885:−1.346

** $p < 0.01$; CI = confidence interval.

4. Discussion

This study aimed to investigate seasonal variation in fat patterning in the rural Ellisras population and to investigate whether there is an association between fat patterning and seasons. There was a significant ($p \leq 0.03$) variation in median fat pattern variables (triceps, biceps, and suprailiac) when comparing by gender in the spring and autumn of 2000. A higher median was observed in girls compared to boys aged 11–15 years. A study reported that girls tend to reach puberty faster than boys [19]. Furthermore, girls are said to gain more fat mass than boys during puberty [20], with faster growth in height, upper arm circumference, and body weight [21]. Therefore, the higher median fat pattern observed among girls in this study may be due to the early onset of puberty among girls. The increase in median fat pattern variables in subscapular and suprailiac values indicate the accumulation of body fat [22]. The increase in median subscapular can be attributed to higher triglycerides and can result in lifestyle-related diseases such as cardiovascular diseases, which are the leading cause of mortality worldwide [23].

There was an insignificant variation in subscapular, triceps, and suprailiac values between spring and autumn among boys aged 5–10 years old. The results differ from those of the previous study which demonstrated a substantial rise in the subscapular skinfold in boys at ages 7, 9, and 10, as well as the suprailiac skinfold at age 9 [22]. Skinfold measurements provide information on changes in body fat distribution and are acknowledged as reliable measures of total body fatness [24], the technique is susceptible to random and systematic errors [25]. Therefore, this explains the difference found in this study and other studies. At all measurement sites, girls exhibit greater median skinfold variables than boys. These results concurred with the findings of the previous studies [26,27]. According to the research finding of the previous study [27], Ellisras girls had considerably larger triceps, biceps, suprailiac, and subscapular skinfold development than boys which is similar to the results from this study.

The seasonal and daily rhythmicity of food intake, metabolism, and adipose tissue function is frequently disrupted in obese people [28]. The prevalence of obesity by ST, SST, and SSTB was mostly significantly higher in spring compared to autumn for both girls and boys, except for boys 11–15 years with respect to the ST ratio. A potential explanation for the higher prevalence in spring might be a decrease in physical activity [29]. The weather has an impact on physical activity [29]. A change in activity patterns from the exterior to the interior entertainment has been linked to an increase in childhood and adolescent obesity [30]. It was reported that in winter there is generally low physical activity and total energy expenditure [28]. The prevalence of obesity ranged from 3–11% with boys being more obese (4–11%) than girls (3–7%) aged 5–15. This contrasted with the observation of the prevalence of overweight among adolescent girls (ages 10–19 years) of the Caboclo of the Amazon basin [31]. The sex steroid hormones, which might be characterised as android patterns for men and gynoid patterns for women, could also have an impact on fat patterning among the Ellisras population. The increase in obesity among boys may be due to variations in sex hormones. A study reported that plasma hormones such as leptin and melatonin influence the development and regulation of obesity. In males, leptin decreases with the increase in sex hormones [28]. In pubertal boys, testosterone is linked to a rise in visceral fat [27]. One study from the same population (Ellisras) showed that obesity prevalence increases with age among children and adolescents [27], with studies from India, Iraq, Mexico, and the United States showing similar results [32–35]. However, this contradicts the results of this study.

The following recommendations may be utilised to interpret correlations: Low = 0.3, moderate = 0.3–0.6, and >0.6 high [36]. These recommendations may suggest a high negative correlation for ST, SST, and SSTB ratios in both autumn and spring. There was a significant negative correlation ($p < 0.001$) in ST, SST, and SSTB ratios in autumn and spring. When two variables are negatively correlated, it signifies that when one increases, the other decreases [37]. The inverse correlation indicated that as seasons change to autumn and spring, the development of obesity decreases. The correlation observed in autumn in this study contradicts those reported by an earlier study which reported an increase in obesity in autumn [38]. The negative correlation observed in spring might be attributed to the fact that spring is the time of the year when physical activity is at its peak [39]. Furthermore, it was reported that during spring there are many outdoor activities such as picnics and physical activity is very high [28].

Limitations of the study include the fact that measurements of plasma insulin, cortisol, melatonin, and leptin could have elucidated some information on the variations of the prevalence of obesity in both girls and boys. Furthermore, the study did not include an analysis of nutritional status, diet, physical activity, and sleep quality, since data collection of these variables was not yet introduced in the ELS. Future studies may investigate seasonal variation in these variables as well. Moreover, the study only investigated autumn and spring in the same year (2000) out of the four seasons known. Therefore, it is not known whether time or seasonality or both played a role in the prevalence of obesity and the correlation between obesity and seasons. Future studies may investigate this using longitudinal data and all four seasons. Although skinfold measurements have been reported to be reliable measures for obesity, the method has limitations in severely obese participants. Therefore, future studies may investigate seasonal variation using other obesity indices.

5. Conclusions

There was seasonal variation among boys and girls in the Ellisras rural population. A higher obesity prevalence was observed in boys rather than girls. Furthermore, a negative correlation between seasons (spring and autumn) and obesity by ST, SST, and SSTB ratio was observed.

Author Contributions: Conceptualisation—K.D.M. and L.M.; Methodology—K.D.M.; Validation and Principal investigator—K.D.M.; Formal analysis—L.M., T.N.M., Investigation—L.M.; Resources—K.D.M.; Collection of data—K.D.M.; Writing—Original Draft Preparation—L.M.; writing—review and editing—L.M., T.N.M., K.D.M., M.M.M., M.M. and S.L.L.; Supervision—K.D.M., M.M. and S.L.L. All authors have read and agreed to the published version of the manuscript.

Funding: This study was funded by the National Research Foundation (grant number: SRUG200303507851).

Institutional Review Board Statement: Ethical approval was granted by the Turfloop research committee (TREC/P/356/2017: PG; date: 2 November 2017).

Informed Consent Statement: All study participants gave their informed consent after being fully informed.

Data Availability Statement: Data will be provided upon request.

Acknowledgments: We would like to sincerely thank the principals of the chosen schools and the field assistants who helped with data collection. We are grateful to the University of Limpopo's Department of Physiology and Environmental Health.

Conflicts of Interest: The authors declare no conflict of interest.

References

1. Gupta, N.; Goel, K.; Shah, P.; Misra, A. Childhood obesity in Developing Countries: Epidemiology, Determinants, and Prevention. *Endocr. Rev.* **2012**, *33*, 48–70. [CrossRef]
2. Toriola, A.L.; Moselakgomo, V.K.; Shaw, B.S. Overweight, obesity and underweight in rural black South African children. *S. Afr. J. Clin. Nutr.* **2012**, *25*, 57–61. [CrossRef]
3. Sedibe, M.H.; Pisa, P.T.; Feeley, A.B.; Pedro, T.M.; Kahn, K.; Norris, S.A. Dietary habits and eating practices and their association with overweight and obesity in rural and urban black South African adolescents. *Nutrients* **2018**, *10*, 145. [CrossRef]
4. Onis, M.; Blossner, M.; Borghi, E. Global prevalence and trends of overweight and obesity among preschool children. *Am. J. Clin. Nutr.* **2010**, *92*, 1257–1264. [CrossRef]
5. Lee, E.Y.; Yoon, K.H. Epidemic obesity in children and adolescents: Risk factors and prevention. *Front. Med.* **2018**, *12*, 658–666. [CrossRef]
6. Li, P.; Yang, F.; Xiong, F.; Huo, T.; Tong, Y.; Yang, S.; Mao, M. Nutritional status and risk factors of overweight and obesity for children aged 9–15 years in Chengdu, Southwest China. *Biomed. Cent. Public Health* **2012**, *12*, 636. [CrossRef] [PubMed]
7. Monyeki, K.D.; Kemper, H.C.G.; Makgae, P.J. Relationship between fat patterns, physical fitness and blood pressure of rural South African children: Ellisras Longitudinal Growth and Health Study. *J. Hum. Hypertens.* **2008**, *22*, 311–319. [CrossRef] [PubMed]
8. Silva, K.S.; Silva, R.C.; Duarte, M.F. Undernutrition and obesity associated with high blood. *Paediatr. Cardiol.* **2009**, *30*, 248–255. [CrossRef]
9. Kemper, H.C.; Post, G.B.; Twisk, J.W.; Mechelen, V. Lifestyle and obesity in adolescence and young adulthood: Results from the Amsterdam Growth and Health Longitudinal Study (AGAHLS). *Int. J. Obes.* **1999**, *23*, S34–S40. [CrossRef]
10. Mitchikpe CE, S.; Dossa RA, M.; Ategbo EA, D.; Raaij JM, A.; Kok, F.J. Seasonal variation in food pattern but not in energy and nutrient intakes of rural Beninese school-aged children. *Public Health Nutr.* **2008**, *12*, 414–422. [CrossRef]
11. Zhu, Z.; Wu, C.; Lou, B.; Zang, J.; Wang, Z.; Guo, C.; Jia, X.; Wang, W.; Shen, X.; Lu, Y.; et al. The dietary intake and its features across four seasons in the Metropolis of China. *J. Nutr. Sci. Vitaminol.* **2019**, *65*, 52–59. [CrossRef] [PubMed]
12. Park, K.S.; Lee, M.G. Effects of summer school participation and psychosocial outcomes on changes in body composition and physical fitness during summer break. *J. Exerc. Nutr. Biochem.* **2015**, *19*, 81–90. [CrossRef] [PubMed]

13. Monyeki, K.D.; Siweya, H.J.; Kemper, H.C.G.; Ramoshaba, N.E. Health Status and Permanent Loss to Follow Up of Ellisras Longitudinal Study Subjects: Rural South African Context. In *Nutrition in Health and Disease—Our Challenges Now and Forthcoming Time*; IntechOpen: London, UK, 2019. [CrossRef]
14. Weather to Visit. What's the Weather Like in Lephalale (Limpopo) in May 2023? 2023. Available online: www.weather2visit.com (accessed on 12 June 2023).
15. Monyeki, K.D.; Cameron, N.; Getz, B. Growth and nutritional status of rural South African children 3–10 years old: The Ellisras growth study. *Am. J. Hum. Biol.* **2000**, *12*, 42–49. [CrossRef]
16. Monyeki, K.D.; Kemper, H.C.G.; Makgae, P.J. The association of fat patterning with blood pressure in rural South African children: The Ellisras Longitudinal Growth and Health Study. *Int. J. Epidemiol.* **2006**, *35*, 114–120. [CrossRef] [PubMed]
17. Norton, K.; Olds, T. *Anthropometrica*; University of New South Wales Press: Sydney, Australia, 1996; pp. 120–267.
18. Dietz, W.H. Critical periods in childhood for the development of obesity. *Am. J. Clin. Nutr.* **1994**, *59*, 955–959. [CrossRef]
19. Pinyerd, B.; Zipf, W.B. Puberty—Timing is everything! *J. Pediatr. Nurs.* **2005**, *20*, 75–82. [CrossRef]
20. Loomba-Albrecht, L.A.; Styne, D.M. Effect of puberty on body composition. *Curr. Opin. Endocrinol. Diabetes Obes.* **2009**, *16*, 10–15. [CrossRef]
21. Thakur, R.; Gautam, R.K. Pre- and post-pubertal growth difference among boys and girls of 5–18 years of age: A cross-sectional study among central Indian Population. *Hum. Biol. Rev.* **2017**, *6*, 164–187.
22. Leal, D.B.; de Assis, M.A.A.; González-Chica, D.A.; da Costa, F.F.; de Andrade, D.F.; Lobo, A.S. Changes in total and central adiposity and body fat distribution among 7–10-year-old schoolchildren in Brazil. *Public Health Nutr.* **2015**, *18*, 2105–2114. [CrossRef]
23. Stacey, N.; van Walbeek, C.; Maboshe, M. Energy drink consumption and marketing in South Africa. *Prev. Med.* **2017**, *105*, 32–36. [CrossRef] [PubMed]
24. Sardinha, L.B.; Going, S.B.; Teixeira, P.J.; Lohman, T.G. Receiver operating characteristic analysis of body mass index, triceps skinfold thickness, and arm girth for obesity screening in children and adolescents. *Am. J. Clin. Nutr.* **1999**, *70*, 1090–1095. [CrossRef]
25. Ulijaszek, S.J.; Kerr, D.A. Anthropometric measurement error and the assessment of nutritional status. *Br. J. Nutr.* **1999**, *82*, 165–177. [CrossRef]
26. Kuhle, S.; Ashley-Martin, J.; Maguire, B.; Hamilton, D.C. Percentile curves for skinfold thickness for Canadian children and youth. *Peer J.* **2016**, *4*, e2247. [CrossRef]
27. Monyeki, K.D.; Kemper, H.C.; Makgae, P.J. Development and tracking of central patterns of subcutaneous fat of rural South African youth: Ellisras longitudinal study. *Biomed. Cent. Pediatr.* **2009**, *9*, 74. [CrossRef]
28. Kanikowska, D.; Sato, M.; Witowski, J. Contribution of daily and seasonal biorhythms to obesity in humans. *Int. J. Biometeorol.* **2015**, *59*, 377–384. [CrossRef] [PubMed]
29. Von Hippel, P.; Benson, R. Obesity and the natural environment across US counties. *Am. J. Public Health* **2014**, *104*, 1287–1293. [CrossRef] [PubMed]
30. Sahoo, K.; Sahoo, B.; Choudhury, A.K.; Sofi, N.Y.; Kumar, R.; Bhadoria, A.S. Childhood obesity: Causes and consequences. *J. Fam. Med. Prim. Care* **2015**, *4*, 187–192. [CrossRef]
31. Silva, H.P.; Veiga, G.V.; Kac, G.; Pereira, R.A. Anthropometric measurements of adolescents from two Amazonian ecosystems: Variations according to seasonality. *J. Biosoc. Sci.* **2010**, *42*, 145–160. [CrossRef]
32. Kumar, S.; Mahabalaraju, D.K.; Anuroopa, M.S. Prevalence of Obesity and Its Influencing Factor among Affluent School Children of Davangere City. *Indian J. Community Med.* **2007**, *32*, 15–17. [CrossRef]
33. Lafta, R.K.; Kadhim, M.J. Childhood obesity in Iraq: Prevalence and possible risk factors. *Ann. Saudi Med.* **2005**, *25*, 389–393. [CrossRef] [PubMed]
34. Perichart-Perera, O.; Balas-Nakash, M.; Schiffman-Selechnik, E.; Barbato-Dosal, A.; Vadillo-Ortega, F. Obesity increases metabolic syndrome risk factors in school-aged children from an urban school in Mexico City. *J. Am. Diet. Assoc.* **2007**, *107*, 81–91. [CrossRef] [PubMed]
35. Skinner, A.C.; Ravanbakht, S.N.; Skelton, J.A.; Perrin, E.M.; Armstrong, S.C. Prevalence of obesity and severe obesity in US children, 1999–2016. *Pediatrics* **2018**, *141*, e20173459. [CrossRef] [PubMed]
36. Marshall, S.J.J.A.; Sark, I.N.; Sallis, J.F. Tracking of health-related fitness components in young ages 9 to 12 years. *Med. Sci. Sports Exerc.* **1998**, *30*, 910–916.
37. Nickolas, S. What Do Correlation Coefficients Positive, Negative, and Zero Mean? Available online: https://www.investopedia.com/ask/answers/032515/what-does-it-mean-if-correlation-coefficient-positive-negative-or-zero.asp (accessed on 25 October 2022).
38. Kobayashi, M.; Kobayashi, M. The relationship between obesity and seasonal variation in body weight among elementary school children in Tokyo. *Econ. Hum. Biol.* **2006**, *4*, 253–261. [CrossRef]
39. Yoshimura, E.; Tajiri, E.; Hatamoto, Y.; Tanaka, S. Changes in season affect body weight, physical activity, food intake, and sleep in female college students: A preliminary study. *Int. J. Environ. Res. Public Health* **2020**, *17*, 8713. [CrossRef]

children

MDPI

Article

Support Needs of Parents of Adolescents Abusing Substances in Selected Hospitals in Limpopo Province †

Lina Sebolaisi Hlahla [1,*], Charity Ngoatle [1] and Tebogo Maria Mothiba [2]

1 Department of Nursing Sciences, University of Limpopo, Polokwane, Sovenga 0727, South Africa
2 Faculty of Health Sciences, University of Limpopo, Polokwane, Sovenga 0727, South Africa
* Correspondence: sebolaisi.hlahla@ul.ac.za

† This is a part of the PhD thesis of Lina Sebolaisi Hlahla in University of Limpopo, Polokwane, RSA.

Abstract: Parents with adolescents who abuse substances need support. They have high stress levels and low quality of life compared to other parents. This is because they have unmet support needs, do not know what to do to help their adolescents, and are distressed. Most studies focus on the support needs of adolescents. Less is known about the specific support needs of their parents. This study explores the support needs of parents of adolescents abusing substances and being treated in five hospitals in Limpopo Province. A qualitative research approach was applied with an explorative, descriptive, and contextual design using semi-structured interviews to understand parents' views. Data saturation was reached at the 14th parent. Data were analyzed using the Tesch method. The parents mainly wanted informational and emotional support. The study identified specific parent-related support needs and adolescent-related support needs. This study is the first to explore the support needs of parents of adolescents abusing substances in a South African rural context.

Keywords: support; parents; adolescent; substances abuse

1. Introduction

Parents with adolescents abusing substances need support. They suffer stress, frustration and anger related to the abuse of substances by their adolescents. This is because parents are afraid that their adolescents may end up suffering mental illness related to substance abuse and later become a burden to the whole family [1]. Parents have a responsibility to manage their adolescents. They do that by shaping the behavior of their children, which is called parental control. That includes setting rules of conduct, supervising the adolescents, and disciplining them. Parents can also show warmth to their children by showing that they are involved in their lives emotionally and physically. Parental warmth and parental control have proven to have protective roles in reducing the abuse of substances by adolescents [2].

When adolescents abuse substances, parents are blamed by society. Parents also blame themselves. They are expected to cope with the consequences of their importance to their children. The results may range from depression and anxiety to humiliation and feelings of loss. They also suffer from guilt, worry, family conflicts, and shame. Parents further fail to cope with the stress and strain caused by the abuse of substances by adolescents [3].

There are negative impacts of adolescent substance abuse within a family. The use of substances by adolescents may lead to family disorganization, disharmony, family disruption, family conflicts, and financial burdens. Parents find adolescents who abuse substances challenging to manage. They go through different emotions, such as shame, guilt, anger, resentment, and depression. They are generally embarrassed or ashamed of their child's substance abuse [4,5].

Parents are affected by the abuse of substances by their children. They suffer in silence with little or no support. Most substance abuse treatment centers treat people with substance abuse dependence, neglecting the families that brought them to the center. There is a need for support services to help parents cope with the challenges they endure as a result of their child's substance abuse [6].

The inability to control adolescents abusing substances leads to parents feeling helpless. It leaves them despondent. They feel hopeless when they have made efforts such as talking to the adolescent about the dangers of taking substances, but their efforts could not yield better results. It is more frustrating for parents when they have taken their adolescents to the rehabilitations centers before but they continue abusing substances [7].

This study explores the support needs of parents of youths who are abusing substances in selected hospitals in Limpopo Province, South Africa. Currently, there are no support programs for the parents of youths who are abusing substances in Limpopo Province. Historically, the use of substances was associated with middle-class communities; however, because of availability of cheaper drugs such as cannabis, the abuse of substances is increasing in rural provinces such as Limpopo [8]. The report by the South African Community Epidemiology Network on Drug Use states that there has been an increase in the number of users seeking treatment for the abuse of substances by young people in the Northern Provinces of South Africa, such as Limpopo [9]. An increase in the abuse of substances by adolescents has put a burden on parents who have to deal with the consequences of the of the addiction of their adolescents on a daily basis [10]. Thus, the there is a need to support parents.

2. Methods

2.1. Research Design

This study adopted a qualitative research approach using an explorative, descriptive, and contextual design. The explorative design assisted in the collection of the views of parents regarding their support needs. An illustrative method was adopted to summarize and synthesize the support needs of parents of adolescents abusing substances in the context of the study, which was a selected hospital in Limpopo Province.

2.2. Study Setting

The study was conducted in the selected hospitals of Limpopo Province. Limpopo Province is located in the northern part of the Republic of South Africa. The reason for the study to be conducted in Limpopo Province is that there has been an increase in the abuse of substances by adolescents in Limpopo Province. The abuse of substances is different based on district. Based on the report from the Department of Social Development, 26% of adolescents in Limpopo Province between the ages of 11 and 20 abuse substances, with the commonly abused substances being cannabis at 49%, inhalants at 34%, and alcohol at 58%. The authors wanted to know the support needs of parents of children abusing substances. Limpopo Province has five districts: Mopani, Sekhukhune, Waterberg, Vhembe, and Capricorn. The regional hospitals used in this study are Letaba, Jane Furse, Mokopane hospital, Tshilidzini, and Mankweng. The wards chosen for this study were those for adolescents admitted because of problems related to substance abuse.

2.3. Population and Sampling

The population for this study was the parents of adolescents abusing substances in selected hospitals in Limpopo Province. The parents of adolescents using substances were accessed during visiting hours and when they brought their adolescents for consultations. Homogeneous purposive sampling strategies were applied to select the parents. This sampling method was used because it ensured that the parents were representative of the population and met the standards the authors desired. They had all the qualities the authors wanted: being the parent of the adolescent abusing substances and staying with the adolescent abusing substances.

2.4. Data Collection Procedures

Data were collected through semi-structured interviews which lasted for 30 to 60 min. The central question that guided the study was, "what are your support needs as the parent of an adolescent abusing substances". As the parents answered the main question, the authors enquired further to obtain more information. In addition to taking field notes while gathering data, the authors used a voice recorder to capture their results. At the 14th parent, data saturation was attained. No incentives were offered to the parents for participation in the study. The 14 parents were either the father or the mother of the admitted adolescent. Table 1 presents the demographics of the parents.

Table 1. Demographics.

Parents	Number of Participants	n = 14
Gender	Female	12
	Males	2
Residential	Rural	12
	Semi-rural	2
	urban	0
Level of education	Primary school	2
	Secondary school	10
	Tertiary	2
Employment	Employed	2
	Unemployed	12

3. Data Analysis

The following eight steps of Tesch's data analysis were used to analyze data: Data were read by authors and converted to codes. The authors re-read the transcripts and analyzed them. There were checks made for duplication of codes and topics. Then, themes and subthemes were created. Data belonging to each theme and subtheme were assembled. The independent co-coder consulted to come to a consensus on themes and subthemes. The existing data were then recorded [11].

3.1. Measures to Ensure Trustworthiness

Measures to ensure trustworthiness were adhered to. Credibility was ensured through prolonged engagement with parents during data collection for two months until data saturation was reached. An audit trail was used to ensure confirmability. To properly understand the topic under research, the authors conducted recorded in-depth interviews and kept field notes. Verbatim transcripts were later written. Dependability was confirmed using an independent coder to reach a consensus regarding themes and subthemes. A full description of the research methodology used in this study assisted in achieving transferability.

3.2. Ethical Clearance and Ethical Considerations

Ethical approval for this study was received from the University of Limpopo's Turfloop Research Ethics Committee (TREC), number: TREC/305/2018: PG. Officials were the Department of Health Limpopo Province (approval number LP 201902_009), the Department of Health district offices in Limpopo Province, and the CEOs from the hospitals selected. The parents were briefed about the study, and agreed to be part of the study voluntarily. The authors let the participants know they had the right to withdraw at any time without repercussions. The researchers were assured of confidentiality and anonymity. Parents signed a consent form before the interview began.

4. Results

The parents' responses during data collection led to the development of the themes and subthemes tabulated below. Table 2 presents the themes and subthemes as they emerged from the study.

Table 2. Themes and subthemes.

Themes	Subthemes
1. Parent-related support needs	1.1 Need for knowledge related to substances and substance abuse
	1.2 Need for feedback
	1.3 Need for support groups
2. Adolescent-related support needs	2.1 Adolescent awareness programs related to substances
	2.2 Support groups for adolescents abusing substances
	2.3 Availability of institutions for adolescents abusing substances

Theme 1: Parent-related support needs
Subtheme 1.1. Need for knowledge related to substances and substance abuse

The study found that the parents of adolescents abusing substances wanted to know about substance abuse. They were failing to cope with their adolescents using substances. They did not know what to do with their adolescents. The parents gave the following accounts describing their need for substance knowledge.

> **Parent 4 said,** *"I wish they could teach me about his substance abuse problem. Maybe they must also teach us their coping strategies with these many patients in the ward. I don't know what to do with him when he is home. I need to learn more about substance abuse. Maybe if there are foods he must not eat, they must tell me about them. I want to learn as much as possible about his condition".*
>
> *(Parent 4, age 45, female)*
>
> **Parent 3 said,** "I need to know if my child is complicating after using substances and how I will help him. What kind of medications can I use for him to stop using substances? Again, how can I support him in his journey of quitting substances because we were not taught how to keep our kids who use substances? As a parent, I know nothing about this substance abuse. I get surprised when I see him high".
>
> (Parent 3, age 39. female)
>
> **Parent 7 said,** "Maybe if the nurses and doctors can teach me how to handle him and help him stop using the substances, I will be a better person".
>
> (Parent 7, 47. male)
>
> **Parent 9 said,** "My family does not know what to do, and I also do not know what to do".
>
> (Parent 9, age 54, female)

Subtheme 1.2: Need for feedback

Parents reported that they needed feedback from healthcare providers. They needed some encouragement when they were doing well. When parents received feedback from healthcare providers, they became confident and less worried. This is supported by the statements below from parents:

Parent 1 said, "They must greet us and tell us what is happening with the patient to make us feel better because I am always worried".

(Parent 1, age 32, female)

Parent 6 said, "If they can just talk to us and reassure us that we are doing a good job taking care of the patient at home, that will be enough".

(Parent 6, age 40, female)

Parent 7 said, "the nurse needs to be kind to us as relatives. They must greet us and tell us what is happening with the patient to make us feel better because I am always worried".

(Parent 7, age 47, male)

Subtheme 1.3: Need for support groups

Parents reported that they needed support groups. This study revealed that parents knew about support groups and the impact they could have on their lives as a form of support. This was confirmed by the parents who gave the views below:

Parent 7 said, "Maybe there can be a training that we attend every week. He must also be part of that training to learn *how to get out of substances*".

(Parent 7, age 47, male)

Parent 11 said, "I heard in Gauteng there are some meetings for the parents of the patients who use substances. We don't have much in this hospital. I wish we had such. That way, we will be quickly involved in the care of my son".

(Parent 11, age 44, female)

Parent 5 said, "I am frustrated I do not know what we are dealing with currently. I can do some support programs for parents. I also wish we had aftercare where these patients are regularly met to see their progress, or the home-based carers must visit us often to check on us. Now have to stop working and come here like I am the one who is sick. I am not doing well".

(Parent 5, age 50, female)

Parent 13 said, "Maybe they can also bring the parents who have been with children who abused substances to come and relate to how they manage their children who use substances. Maybe they can give us coping strategies".

(parent13, age 48, female)

Theme 2: Adolescent-related support needs.

Parents reported that the following factors could support them as they cared for adolescents abusing substances.

Subtheme 2.1: Adolescents awareness programs related to substances.

Parents wished that there could be awareness programs for adolescents abusing substances. Awareness programs serve as a way of sharing information about the dangers of substances to the adolescents. The claims below serve as evidence of what the parents said:

Parent 14, "I wish they could bring a person who was once a substance abuser who can come and talk to my son about how he got out of substances".

(Parent 14, age 48, female)

Parent 6, "I hope there could be programs for drug users like alcohol anonymous for alcoholics".

(Parent 6, age 37, female)

Parent 9 said, "They must also teach my son to accept himself. He is always isolated".

(Parent 9, 54)

Parent 2, "I wish they could talk to my son and tell him that substances are unsuitable and negatively affect his bright future."

(Parent 2, age 52, female)

Subtheme: 2.2: Support groups for adolescents abusing substances

Just as parents want support groups for themselves as the parents of adolescents abusing substances, they also suggested the need for support programs for adolescents abusing substances. The quotes below support their need for support programs:

> **Parent 9 said,** "Maybe if we can get programs regarding the adolescent who are using substances ... the substances are busy destroying their bodies".
>
> (Parent 9. Age 54, female)

> **Parent 6 said**, "I also wish we had aftercare where these patients are regularly being met to see their progress, or the home-based carers must visit us often to check on us. Now I had to stop working and come here like I was the one who was sick. I am not doing well".
>
> (Parent 6, age 37, female)

> **Parent 7 said,** "There must also be community centers or programs that help during discharge. This thing takes the whole life to be truly liberated. At the hospital, we get help, but when we go home, nothing happens".
>
> (parent 7, age 47, male)

Subtheme 2.3: Availability of institutions for adolescents abusing substances

Parents reported that the availability of institutions could serve as a support to them. This is because rehabilitation centers provide care and treatment for longer periods where adolescents will be separated from friends who might influence their substance abuse. This is supported by the following statements from study participants:

> **Parent 14 said,** "I also hope they can refer us to the rehabilitation center. I do not know if the government can make such referrals to go there for some time. He must be away for some time because his friends are waiting for him when he returns home. He will soon go back to substance abuse. All I want is for him to be okay. I do not care how long he goes away".
>
> (Parent 14, age 42, female)

> **Parent 1 said,** "I wish the hospital could open a rehabilitation center or maybe write a letter for us to take him to the rehabilitation center. They must keep him in the hospital for a long time. They can also refer us to a psychologist because his substance abuse does not affect him alone. It affects us all as a family".
>
> (Parent 1, age 32, female)

> **Parent 12 said,** "There must also be community centers or programs that help during discharge. This takes the whole life to be truly liberated. At the hospital, we get help, but when we go home, nothing happens".
>
> (Parent 12, age 60, female)

5. Discussion

This study aimed to explore the support needs of parents of adolescents abusing substances. There are two main themes that emerged from the study. The themes were parent-related support needs, which had three subthemes, and adolescent-related support, which had three subthemes.

The results indicate that parents need to know more about substances. Parental knowledge regarding substance abuse plays a role in reducing the abuse of substances by adolescents. When parents know about the abuse of substances, it is easy for them to discipline their children and communicate with them about the dangers of substance abuse. With high parental knowledge, there are low rates of substance abuse [2]. Obtaining information from healthcare providers is the right of parents and of adolescents abusing substances. It is, therefore, the duty of healthcare providers to share information with parents at every opportunity they have to be with the parents. The need for information by parents includes being given opportunities to ask questions and being allowed to be part of the decision-making process around the care of their adolescents [12–14]. The anxieties that parents have can be reduced when parents know about substance abuse. Healthcare providers should therefore use every opportunity they have to share information with parents.

The results of this study indicate that parents of adolescents abusing substances need feedback. They want to know whether they are doing well or not taking care of their adolescents using substances. This is supported by Davis-Strauss et al. [15], who stated that when healthcare professionals provide feedback to the parents, they make them feel safe, increasing their belief that parents want more coordinated and individualized care as well as to be reassured by the healthcare professionals in their parenting and caregiving during the transition of assuming the responsibility of taking care of the adolescents abusing substances. Parents prefer support from healthcare providers even though they have family support. Healthcare providers must make a conscious effort to show concern for parents and patients. Healthcare providers must conduct support visits to parents to encourage them to keep doing well. According to Loft et al. [16], a good relationship between parents and healthcare professionals improves the outcomes of care by the parents. Healthcare professionals need to show empathy to parents. They need to be able to talk to parents, encourage them to do well, and guide them where they fail so that they avoid the stigma related to being a parent of a substance abuser.

The parents, as participants in this study, showed that they need support groups. Support group interventions can assist parents in obtaining social support from fellow parents who undergo similar situations. Support groups provide peer support with the help of a facilitator who can be a healthcare professional. It can also occur when parents are matched with other parents going through the same problem as caring for the youth who are abusing substances [17]. This is because, according to Caldarera et al. [18], support groups are a valuable tool for working closely with parents. The participation of parents in support groups gives parents a better understanding of their situations. They can understand their emotional challenges and their children's challenges. Support groups have also been found to improve parent–child relationships. With support groups, parents can gain confidence and feel less lonely. It also enhances the well-being of parents by giving them optimistic hope. The primary purpose of support groups is to encourage parents to share their consent, knowledge, and achievements in a mutually supportive network. The study by Jackson et al. [19] reported that support groups can be helpful for the parents of youth who are abusing substances. However, the support groups' design and content are essential to the parents. It is necessary for the healthcare workers to conduct the support group in such a way that they are helpful to parents. Support groups can be done virtually and in person. An online support group is an alternative for parents, as many parents go online to seek information and support [20].

The study results have indicated that parents want to be supported through awareness programs for youth who are abusing substances. Adolescents are reluctant to seek help for several reasons, such as a lack of knowledge regarding substance abuse, poor understanding of the dangers of substances, and where to find help if addicted. They believe they can handle their problems. The other factor can be previous bad experiences [15]. There is a need for substance abuse awareness campaigns in communities to enlighten adolescents about the adverse effects of substance abuse. Awareness programs will capacitate adolescents with all the information they need regarding substances [21]. Awareness programs must include factors to increase youth involvement and participation. When awareness programs encourage youth involvement, they build relevance for youth and can eliminate many obstacles that may make them uninterested in such programs. Part of n awareness program's content should be to make communities aware that their community beliefs and norms can contribute to the abuse of substances. Communities should be mindful of the effects of standards and beliefs, especially if they lead to substance abuse.

Substance abuse awareness campaigns in congregations will enlighten the community about the adverse impact of substance abuse [15,21]. The awareness program can assist in the spread of knowledge regarding substance abuse to adolescents and slow down the abuse of substances by the adolescents.

The study findings revealed that parents want a space where adolescents can meet and share their experiences with substance abuse. There is a need for support groups for adolescents. These adolescent support groups can serve as a support to the parents. According to Sellers et al. [22], peer support has the potential to enhance recovery. The outcomes can improve outcomes for both the youth and their families. Support groups are associated with high self-esteem among youth, low anxiety and depression, and better sleep. Support is positively related to improved mental and physical health [23]. Peer support groups may be able to provide support to youth who are abusing substances and can fulfill the function of informational and emotional support. Peer support can alleviate the youth's mental health stress and provide social support. It can also motivate the youth to take better care of themselves. There will be better coping mechanisms among the youth [24]. Support groups can assist the youth to better take care of themselves and decrease the likelihood of relapse. As support groups help with coping mechanisms, they can also assist with buffering against the stigma related to substance abuse and its consequences.

The study revealed that parents want institutions where they can send their adolescents for further treatment and rehabilitation. An alternative approach to dealing with substance abuse is required. Establishing rehabilitation centers that can be accessed by all members of society from different socioeconomic statuses is necessary. Accessibility to such a center can help reduce the abuse of substances by the youth and serve as a support system for the youth. The advantage of rehabilitation centers is that they provide multidisciplinary assessments for creating an individualized plan [25,26]. According to Mohasoa and Mokoena [27], there is a need for services that can help with aftercare for substance abuse. Aftercare services are associated with reduced relapses, and they equip substance abusers with skills to maintain treatment gains and sobriety.

Based on the discussions provided, it is recommended that parents be supported, and the support should be offered from healthcare institutions—healthcare professionals need to ensure that the parents of admitted adolescents abusing substances are given attention. Healthcare professionals should create a time to talk about how they are doing as parents of youth using substances. Healthcare providers must use every opportunity they have to give the parents information. Information can either be on substance abuse or the progress of their adolescent in the hospital. If healthcare providers have no time to communicate with the parents due to workload, they can make sure that within hospitals there are posters and information boards addressing substance abuse issues. Within the wards, where parents visit their adolescents, there must be information pamphlets addressing substance abuse.

It is also recommended that health institutions develop programs where the parents are educated on substances and substance abuse in a formal way. The agenda should cover the parents' support needs, including sharing information regarding substance abuse. These kinds of programs can start from community clinics through to health education and awareness campaigns to prevent hospital admissions. The healthcare professional can support parents by organizing support groups in which parents will have a safe space to share their experiences in dealing with adolescents abusing substances.

6. Limitations

This study was conducted at selected hospitals in Limpopo Province, so the study's findings cannot be generalized to other settings.

7. Conclusions

The study investigated the support needs of parents of adolescents abusing substances. Parents indicated that they could feel supported when given information regarding the implications; they also wanted to know if what they were doing in taking care of their adolescents abusing substances was correct; and they also need support groups. They also gave an account of how they want to be supported through the care they wanted their adolescents to receive. Healthcare professionals need to focus on parents as much as they care for adolescents abusing substances. There should be information sessions between parents and the healthcare providers. The development of support groups for parents and adolescents using substances can assist in reducing the pressure the parent goes through when taking care of the adolescents abusing substances. Implementing the suggested support needs of parents will assist in managing the stress and anxiety of the parents who are managing their adolescents abusing substances.

Author Contributions: Conceptualization, L.S.H. and T.M.M.; methodology, L.S.H.; validation, C.N.; formal analysis and C.N.; investigation, L.S.H. and T.M.M.; resources, T.M.M. and C.N.; data curation, L.S.H.; writing—original draft preparation, L.S.H.; writing—review and editing, L.S.H., C.N. and T.M.M.; visualization, L.S.H.; supervision, T.M.M. All authors have read and agreed to the published version of the manuscript.

Funding: This research received no external funding.

Institutional Review Board: The study was conducted in accordance with the Declaration of Helsinki, and approved by the Institutional Review Board (or Ethics Committee) of Turfloop Research Ethics Committee (TREC/305/2018:PG, approval on 27 November 2018, University of Limpopo).

Informed Consent Statement: Informed consent was obtained from all parents involved in the study.

Data Availability Statement: Not applicable.

Acknowledgments: This study was supported by the Tirisano Track A project (UCLA).

Conflicts of Interest: The authors declare no conflict of interest.

References

1. Collins, A.; Burchell, J.; Remedios, C.; Thomas, K. Describing the psychosocial profile and unmet support needs of parents caring for a child with a life-limiting condition: A cross-sectional study of caregiver-reported outcomes. *Palliat. Med.* **2020**, *34*, 358–366. [CrossRef] [PubMed]
2. Trucco, E.M. A review of psychosocial factors linked to adolescent substance use. *Pharmacol. Biochem. Behav.* **2020**, *196*, 172969. [CrossRef] [PubMed]
3. Groenewald, C.; Bhana, A. Mothers' experiences of coping with adolescent substance abuse: A phenomenological inquiry. *Contemp. Nurse* **2017**, *53*, 421–435. [CrossRef] [PubMed]
4. Subekti, H.; Wilopo, S.A.; Rahmat, I. Stress adaptation among family of adolescent with substance use disorders Systematic literature review. *Open Access Maced. J. Med. Sci.* **2021**, *9*, 335–341. [CrossRef]
5. Saquib, N. Substance use disorder: A growing but understudied mental health condition. *Int. J. Health Sci.* **2019**, *13*. 1–2.
6. Groenewald, C.; Bhana, A. Substance abuse and the family: An examination of the South African policy context. *Drugs Educ. Prev. Policy* **2018**, *25*, 148–155. [CrossRef]
7. Aoun, S.M.; Gill, F.J.; Phillips, M.B.; Momber, S.; Cuddeford, L.; Deleuil, R.; Stegmann, R.; Howting, D.; Lyon, M.E. The profile and support needs of parents in paediatric palliative care: Comparing cancer and non-cancer groups. *Palliat. Care Soc. Pract.* **2020**, *14*, 1–14. [CrossRef]
8. Dada, S.; Burnhams, N.H.; Erasmus, J.; Parry, C.; Bhana, A. *South African Community Epidemiology Network of Drug Use (SACENDU): Monitoring Alcohol, Tobacco and Other Drug Abuse Treatment Admissions in South Africa*; SAMRC: Cape Town, South Africa, 2018. Available online: https://www.samrc.ac.za/sites/default/files/attachments/2019-10-16/SACENDUFullReportPhase45.pdf (accessed on 10 March 2023).
9. Hornsby, N.; Harker, N.; Jodilee, B.; Warren, E.; Siphokazi, L.; Parry, C.; Bhana, A.; Pretorius, S.; Weimann, R. South African Community Epidemiology Network of Drug Use (SACENDU): January–June 2021 phase 50. Monitoring alcohol, tobacco and other drug abuse, Treatment admissions in south africa. 2022. Available online: https://www.samrc.ac.za/intramural-research-units/atod-sacendu (accessed on 10 March 2023).
10. Hlungwani, E.N.; Ntshingila, N.; Poggenpoel, M.; Myburgh, C.P.H. Experiences of parents with an adolescent abusing substances admitted to a mental health institution in Giyani, South Africa. *Curationis* **2020**, *43*, 1–9. [CrossRef]
11. Creswell, J. *Research Design: Qualitative, Quantitative, and Mixed Methods Approaches—John W. Creswell—Google Books*; Megan Markanich, M., Ed.; Fourth; SAGE Publications, Inc.: Los Angeles, CA, USA, 2014; ISBN 978-1-4522-2609-5.
12. Birtel, M.D.; Wood, L.; Kempa, N.J. Stigma and social support in substance abuse: Implications for mental health and well-being. *Psychiatry Res.* **2017**, *252*, 1–8. [CrossRef]

13. Leow, M.Q.H.; Chan, S.W.C. The challenges, emotions, coping, and gains of family caregivers caring for patients with advanced cancer in Singapore: A qualitative study. *Cancer Nurs.* **2017**, *40*, 22–30. [CrossRef]
14. Saria, V.F.; Mselle, L.T.; Siceloff, B.A. Parents and nurses telling their stories: The perceived needs of parents caring for critically ill children at the Kilimanjaro Christian Medical Centre in Tanzania. *BMC Nurs.* **2019**, *18*, 1–8. [CrossRef] [PubMed]
15. Davis-Strauss, S.L.; Johnson, E.; Lubbe, W. Information and Support Needs of Parents With Premature Infants: An Integrative Review. *J. Early Interv.* **2021**, *43*, 199–220. [CrossRef]
16. Loft, M.I.; Martinsen, B.; Esbensen, B.A.; Mathiesen, L.L.; Iversen, H.K.; Poulsen, I. Call for human contact and support: An interview study exploring patients' experiences with inpatient stroke rehabilitation and their perception of nurses' and nurse assistants' roles and functions. *Disabil. Rehabil.* **2019**, *41*, 396–404. [CrossRef] [PubMed]
17. Sartore, G.M.; Pourliakas, A.; Lagioia, V. Peer support interventions for parents and carers of children with complex needs. *Cochrane Database Syst. Rev.* **2021**, *2021*, 12. [CrossRef]
18. Caldarera, A.M.; Davidson, S.; Vitiello, B.; Baietto, C. A psychological support group for parents in the care of families with gender diverse children and adolescents. *Clin. Child Psychol. Psychiatry* **2021**, *26*, 64–78. [CrossRef]
19. Jackson, C.; Cheater, F.M.; Reid, I. A systematic review of decision support needs of parents making child health decisions. *Health Expect.* **2008**, *11*, 232–251. [CrossRef]
20. Friedman, E.M.; Trail, T.E.; Vaughan, C.A.; Tanielian, T. Online peer support groups for family caregivers: Are they reaching the caregivers with the greatest needs? *J. Am. Med. Inform. Assoc.* **2018**, *25*, 1130–1136. [CrossRef]
21. Rose, C.; Peter, M.; Samwel, O. an Investigation of the Factors Contributing To Drug and Substance Abuse Among the Youth in Kenya: A Survey of Select Rehabilitation Centres in Mombasa County. *Int. J. Public Health* **2019**, *1*, 53–70.
22. Sellers, C.M.; O'Brien, K.H.M.; Hernandez, L.; Spirito, A. Adolescent alcohol use: The effects of parental knowledge, peer substance use, and peer tolerance of use. *J. Soc. Social Work Res.* **2018**, *9*, 69–87. [CrossRef]
23. Zarshenas, L.; Baneshi, M.; Sharif, F.; Moghimi Sarani, E. Anger management in substance abuse based on cognitive behavioral therapy: An interventional study. *BMC Psychiatry* **2017**, *17*, 1–5. [CrossRef]
24. Hameed Shalaby, R.A.; Agyapong, V.I.O. Peer support in mental health: Literature review. *JMIR Ment. Health* **2020**, *7*, 6. [CrossRef]
25. Ghazal, P. Rising trend of substance abuse in Pakistan: A study of sociodemographic profiles of patients admitted to rehabilitation centres. *Public Health* **2019**, *167*, 34–37. [CrossRef] [PubMed]
26. Al Ghaferi, H.A.; Ali, A.Y.; Gawad, T.A.; Wanigaratne, S. Developing substance misuse services in United Arab Emirates: The National Rehabilitation Centre experience. *BJPsych. Int.* **2017**, *14*, 92–96. [CrossRef] [PubMed]
27. Mohasoa, I.; Mokoena, S. Challenges Facing Rural Communities in Accessing Substance Abuse Treatment. *Int. J. Soc. Sci. Humanit. Stud.* **2019**, *11*, 35–50.

children

MDPI

Article

Attitudes of Caregivers of Children under Five Years Regarding Growth Monitoring and Promotion in Polokwane, Limpopo Province

Mabitsela Hezekiel Mphasha *, Matjie Rapetsoa, Nhaviso Mathebula, Kamogelo Makua and Sanele Mazibuko

Department of Human Nutrition and Dietetics, University of Limpopo, Sovenga 0727, South Africa
* Correspondence: pitso85@gmail.com

Abstract: Growth monitoring and promotion (GMP) is critical in tracking child growth to address widespread malnutrition and health status. Attitudes influence behaviour change, including attendance of GMP, and negative attitudes are linked to non-attendance. Moreover, negative attitudes correlate with low socioeconomic position. South Africa is characterized by inequality, which may lead to negative attitudes towards GMP among caregivers with a poor socioeconomic status. Hence, this study seeks to explore the attitudes of caregivers of children under five towards GMP. A qualitative exploratory study design was used. Caregivers of children under five were purposively sampled. Twenty-three participants were interviewed one-on-one, and the data were recorded using voice recorders and field notes. Tesch's eight steps and inductive, descriptive, and open coding techniques were used to analyse the data. Participants understood the significance of GMP and were confident their children would benefit from it; hence, they attended sessions out of love for their children. The inconsistent availability of GMP services and the behaviour of health workers affected participants' attitude. Despite these challenges, participants felt good about GMP. Caregivers' love for their children/grandchildren helped them overcome challenges experienced at the health facilities. Good feelings about GMP boosted caregivers' attitudes and aided in adherence. An intervention to address element impacting attitudes of caregivers is recommended.

Keywords: growth monitoring and promotion; attitudes; caregivers; attendance; children under five

1. Introduction

Malnutrition in children is a serious public health issue that harms health and hinders development [1]. Malnutrition is a factor in more than half of all deaths in children under five years because it puts children at risk for developing and dying from infections. There were 144 million stunted, 47 million wasted, 14.3 million severely wasted, and 38.3 million obese children under the age of five years old in the world in 2019 [2,3]. In sub-Saharan Africa, there were 57.5 million stunted, 11.8 million wasted, 3 million severely wasted, and 5.2 million obese children under the age of five years old in 2019 [1]. South Africa (SA) experiences a triple burden of malnutrition—undernutrition, overnutrition, and micronutrient deficiencies [4,5]. Stunted growth is the most prevalent form of malnutrition, and 80% of the world's stunted children reside in SA [6]. In South Africa, 27% of children under the age of five were stunted, 44% were vitamin deficient, 13% were overweight, and 6% were underweight in 2016 [5]. Malnutrition incidents vary by socioeconomic categories and geographic areas in SA [5]. Growth monitoring and promotion (GMP) is one of the methods that have been used in developing and developed nations to lessen childhood malnutrition [1].

GMP is a nutritional intervention for children under the age of five that involves routinely measuring and plotting their weight on a growth chart at predetermined intervals [7]. A reference group's growth curves are also compared as part of this process. The findings are utilized to guide caregivers to implement specific actions that will enhance the child's growth [8,9]. GMP identifies and addresses growth faltering and provides an early warning signal for the implementation of suitable and timely actions before a child's nutritional status deteriorates into a severe case of malnutrition [1,10]. Immunization, diet, and/or vitamin supplementation, growth monitoring, nutrition education, and counselling are all included in GMP services [11]. GMP has proven to be capable of preserving and improving child-feeding and childcare procedures [12,13]. Attending the GMP sessions allows caregivers to, among other things, speak with healthcare professionals about childcare and receive assurances about the wellbeing of their children [14]. It is important to assess the caregivers' satisfaction. The level of satisfaction and attitudes towards GMP services are correlated.

According to Vargas-Sánchez, [15] attitude is the way a person perceives and assesses something or someone, as well as the tendency to react favourably or unfavourably to specific concepts, things, people, or circumstances. The positive or negative actions that a person has experienced may influence their attitude. In this context, caregivers' good attitudes towards GMP services may aid in honoring follow-up appointments. Amid increasing malnutrition among children, it is, therefore, significant to understand the attitudes of the caregivers of children under five years. Muchiri, Gericke, and Rheeder [16] cite attitudes as one of the key factors influencing behaviour change, which, in this context, may influence caregivers to bring their children to GMP sessions. Passive or negative attitudes were found to be linked to low socioeconomic status. [14] Inequality affects people's viewpoints on things, including involvement in social and health programmes. Researchers have found that those with a lower socioeconomic status experience more stressors, which contribute to a greater number of compulsive behaviours [17,18]. South Africa is characterized by inequality, which may lead to negative attitude towards GMP among caregivers with a poor socioeconomic status, particularly considering that most citizens are poor. This study explores attitudes of caregivers of children under five years regarding growth monitoring and promotion in Polokwane, Limpopo Province.

2. Research Methodology

2.1. Research Method and Design

This study was conducted using a qualitative approach, which followed phenomenological exploratory study design. Caregivers of younger children under five years described their attitudes towards GMP, and findings are captured in the Results section.

2.2. Study Site

This study was conducted in the rural part of Polokwane Municipality, specifically Sebayeng Clinic. Sebayeng is located within the Solomondale area and falls under Dikgale cluster of Polokwane Municipality. The study site was selected to understand challenges related to non-adherence to GMP sessions. GMP services are offered twice a week at this clinic, mainly by primary healthcare nurses and dietitians. Students from other departments such as nursing and human nutrition and dietetics at the University of Limpopo complete their practicums at this clinic by, among other services, offering GMP services.

2.3. Ethical Considerations

The study's use of human volunteers complied with all required ethical guidelines. The Turfloop Research Ethical Committee (TREC) approved the research and issued clearance certificate number TREC/238/2021:UG. The Limpopo Department of Health (DOH) and the clinic manager gave their permission for the study's execution. Participants attested to their free will to participate by signing informed consent form. The confidentiality and privacy of the participant's data were likewise respected.

2.4. Sampling and Participants

The target population for this study were caregivers attending GMP services in Sebayeng Clinic. Caregivers in this context refers to mothers and/or legal guardians who assume the primary responsibility for looking after children under the age of five years. As a result, this clinic was chosen because it contains many caregivers of children under five, which made it simpler to find participants who meet the study's inclusion criteria. Inclusion criteria in this study were only caregivers whose children are enrolled and on the register of GMP programme at Sebayeng Clinic. Twenty-three participants were purposively sampled from the target population of this study. The sample size was established using data saturation.

2.5. Data Collection

Unstructured interviews were conducted in which participants were interviewed one-on-one. The interviews were recorded using voice recorders, and, to prevent recording unintentional off-the-record remarks, participants were informed when the recorders were on and off. The interviews were conducted by student dietitians doing their community nutrition practicums at the selected clinic. The student dietitians were provided with training on how to conduct one-on-one interviews and ask probing questions from participants. Furthermore, pilot study was conducted in the presence of supervisor to assist in improving the skills of the student dietitians. The participants and results of pilot study were not included in the overall study. The interviews were conducted in Sepedi because the people in the area are Sepedi speakers. The main question that was asked to participants at the opening of interviews was ***'What is your attitude towards growth monitoring and promotion?'*** An interview guide was used during interviews; however, follow-up questions or probing was completed.

2.6. Trustworthiness

Trustworthiness was ensured through credibility, transferability, confirmability, and dependability as described in Table 1.

Table 1. Measures of trustworthiness.

Strategy	Criterion	Applicability
Credibility	Prolonged engagement	Researchers responsible for data collections were student dietitians conducting community nutrition practicals at Sebayeng Clinic. Their practicals included offering GMP services to caregivers. Therefore, researchers were familiar with and known in the area. In addition, researchers collected data from each participant for about 15–45 minutes to extract more information.
	Member checks	Immediately after the conclusion of each interview, a summary was provided to the participant for confirmation of findings.
Transferability	Data saturation	Participant number 20 added no new data; an extra 3 participants were further sampled for confirmation of data saturation. No new information was added; data collection was, therefore, discontinued.
	Thick description	The study's setting, sampling procedures, and data collection methods were all clearly explained.

Table 1. *Cont.*

Strategy	Criterion	Applicability
Confirmability	Peer review	Three researchers collected and analysed the data independently, developed themes and sub-themes independently, and met for consensus. Thereafter, they met with supervisors and agreed on themes and sub-themes.
	Reflexivity	The researchers remained neutral during the process of data collections and used probing and reflexivity to acquire more data from participants by using statements such as *"so what you are actually saying is..."*
Dependability	Dense description of research methods	The researchers depended on clear and detailed description of research methods, including the use of participants' quotations in presenting results.
	Audit trail	Researchers who collected the data described to supervisors how the data were gathered and analysed. Themes and sub-themes were agreed upon. Data to be kept in a lockable file for the ensuing five years.
	Reliance on data collection tools, and supervisors	Researchers relied on voice recorder to capture interviews and experiences of supervisors.

Source: Ramathuba and Ndou [19].

2.7. Data Analysis

Each interview was audio recorded, followed by transcription. Prior to analysis, the researchers had the interviews, which were conducted in Sepedi, translated into English by a language translator. The verbatim transcripts were then independently analysed by each researcher and their supervisors. In a consensus meeting, all researchers and an independent coder came to an understanding on the themes and sub-themes that arose from the independent analysis of the transcripts of the interviews. Direct quotations from participants were taken and recorded to corroborate the conclusions. The eight steps of Tesch's open coding qualitative data analysis method by Creswell [20] was used to analyse the data. Researchers carefully and repeatedly read all transcripts and developed codes. Topics from codes were grouped together according to similarity, abbreviated, and written next to the appropriate segments of the transcription. Thereafter, themes and sub-themes were developed.

3. Results

3.1. Demographic Information

The demographic information describes the participants that were interviewed in this study.

Table 2 shows that all 23 participants were females, 16 were between 18 and 35 years and had a high school education, and 13 were unemployed. In addition, 17 of them were married, and their children were between the ages of 1 year and 30 months old, while 11 of the children were brought by their mothers.

Table 2. Socio-demographic profile of participants.

Socio-Demographic Profile	Category	N = 23
Gender	Male	0
	Female	23
Age	18–35	16
	36 or above	7
Marital status	Married	17
	Single	6
Employment	Employed	10
	Unemployed	13
Education	High school education or less	16
	Tertiary education	7
Child's age	1–30 months	17
	31–72 months	6
Relationship to child	Mother	11
	Grandmother	10
	Sibling of mother	2

3.2. Themes and Sub-Themes

Three themes emerged from this study, which are perceptions on GMP, healthcare services, and adherence to sessions. Themes 1 and 2 had two sub-themes each, while theme 3 had one sub-theme.

3.2.1. Theme 1: Perceptions on GMP

Three essential components of perception are cognition, understanding, and the act of perceiving or comprehending through the senses or intellect. GMP frequently acts as a crucial ongoing link between healthcare workers and a child's caregivers, by using a growth chart and counselling. The attitudes and believes of laypeople towards GMP are influenced by personal experience. The study's findings revealed how people view the GMP programme and how well they comprehend GMP. From this theme, the following sub-themes emerged:

Sub-Theme 1.1: Importance of GMP

Understanding and appreciating GMP's worth or significance could help caregivers develop a positive attitude for continuous use of the services. The value of GMP in relation to childcare and health was demonstrated by study participants. The quotes listed below support this sub-theme:

"The GMP is vital for child growth and health; it helps identify children with growth faltering for corrective measures and developmental milestones, so that immediate interventions can be initiated" said **participant 02**.

Participant 21: *"I am aware of the significance of bringing the child to the clinic, particularly in order to prevent infections, chicken pox, diarrhea, and vitamin deficiencies."*

Sub-Theme 1.2: Feelings towards GMP

Negative feelings towards GMP may lead to skipping regular clinic visits. On the other hand, feeling good might make a person use the services regularly. Participants in this study expressed happiness or feeling good, as seen by the following quotes:

Participant 08:*"Despite difficulties such as inconsistent availability of GMP services at the clinics, I am ready to keep attending the regular sessions. I'm pleased with this program. It benefits our kids."*

"I'll advise new or young mothers to bring their kids to the GMP sessions. The kids benefit from this. We must put the difficulties we face in the clinic behind us, and continue to bring our children for these services" said **participant 12.**

3.2.2. Theme 2: Health Services

Healthcare professionals usually deliver GMP services to the caregivers. The availability of services and access to them are essential for adherence. According to the following sub-themes, caregiver attitudes were apparently influenced by the irregular availability of GMP services and the behaviours of the healthcare professionals:

Sub-Theme 2.1: Inconsistent Availability of GMP Services Impacts Attitudes of Caregivers

The general public's constant access to and the availability of healthcare services is essential for enhancing the value of GMP services, including attitudes. In contrast, irregular or inconsistent provision of GMP services may result in caregivers not complying, devaluing the programme, and adopting unfavourable attitudes. As evidence, consider the following claims:

"We occasionally visit clinics for GMP without assistance on account of unavailability of services, which makes us consider how important the GMP, as promoted, is. Our attitudes and perceptions of the GMP, particularly adherence, are impacted by this" said **participant 04**.

Participant 13:*"Sometimes we are told that the clinic's weighing scale is broken or that there are no child immunizations or supplements, so we end up skipping appointments."*

Sub-theme 2.2: Behaviour of Healthcare Hroviders and Attitudes of Caregivers

The healthcare providers who are the drivers or implementers of GMP services to the communities are crucial and may influence caregivers' adherence to sessions. Additionally, the behaviour of these providers may influence the attitudes of the caregivers on how they perceive GMP. This is supported by the following statement from one of the study's interviewees:

"The nurses sometimes speak to us with disrespect, as if we chose for our children not to grow well; the disrespect demotivates us from adhering to sessions. Truly, we turn to develop bad attitudes towards the clinic services including GMP" said **participant 08**.

3.2.3. Theme 3: Adherence to Sessions

Adherence to the GMP sessions is critical to the health and development of the children, who are usually brought to the healthcare facilities by their caregivers. On the other hand, adherence may be impacted by a number of factors, such as the actions of healthcare workers and the irregular provision of services at the clinics. The following sub-theme shows that caregivers responded that they continue to bring their children to the GMP sessions despite these difficulties or factors out of a sense of love:

Sub-Theme 3.1: Attendance of Sessions for the Love of Children

Attending the GMP sessions is essential, since it enables early growth changes in the child to be identified. The growth trends are analysed, and the necessary steps are taken. Participants stated that they try to routinely attend the GMP sessions for their children's sake. The quotations that follow support this:

"I love my child and want him to develop into a healthy adult. I go to the regular clinic appointments on time so that I can find out how my child is doing and provide better care" said **participant 10**.

"I routinely attend the sessions as scheduled because I want my granddaughter to grow up healthy like my children and free from diseases as well" said **participant 15.**

4. Discussions

The key to influencing behaviour is attitude. When they feel good about the programme or have a positive attitude towards it, the caregivers of children under the age of five are more likely to attend the GMP sessions. Therefore, the purpose of this study was to determine attitudes towards GMP among caregivers of children under five in Polokwane, Limpopo Province. The study showed that participants are aware of the significance of GMP. Various studies found that knowledge is related to attitudes as well as behaviour [21,22]; hence, knowledge emerged during the exploration of attitudes. However, the relationships between the knowledge and attitudes of these caregivers should be determined through a quantitative cross-sectional study, where statistical analysis shall be used to determine associations. Studies have shown a correlation between a knowledge of GMP and higher attendance in the programme, and, conversely, a lack of knowledge was linked to low programme attendance. These researchers showed that caregivers are more willing to participate in the GMP activities on a regular basis when they see them as vital and understand them [23,24]. Knowledge is not the only factor; limited community interaction and inadequate child nutrition counselling, such as from the GMP programme, were found to be the main reasons for low attendance, according to an Ethiopian study that examined reasons for non-attendance [25]. Caregivers desire healthy growth and development, including adequate nutrition, for their children. However, some factors such as availability and affordability may impact feeding. An Australian study reported that low socioeconomic status is linked to a lower intake of healthy food choices. It further proposed the need for policy action to improve the affordability of healthy foods [26]. Participants in this study made references to doing it out of love for their children. Even in the case of the irregular availability of these services and/or unwelcoming behaviour or attitudes of healthcare staff, caregivers may be motivated by love for their own children to adhere to the GMP sessions. Love has the capacity to profoundly alter how people approach their pursuit of health. People who believe they have a health problem or are ill are deemed to be engaging in health-seeking behaviour when they take any activity with the goal of finding a remedy that will work for them [27]. By engaging in appropriate healthcare-seeking behaviour, mothers can either prevent or considerably reduce the likelihood of child mortality brought on by childhood illnesses [28–30].

The interviewees in this study made it clear that the inconsistent availability of GMP services and the actions of the healthcare professionals who are the programme's driving forces have an impact on their perceptions or attitudes regarding this programme. GMP services, such as immunization, supplementation, growth monitoring, nutrition education, and counselling, are carried out by health professionals. According to a cross-sectional survey, caregivers complained that healthcare professionals failed to involve them in growth-monitoring processes and provided generally inadequate GMP counselling [1]. Another study found that after weighing sessions, healthcare professionals did not talk with caregivers about their children's growth and development and did not ask them to interpret the growth chart [8]. Along with providing a service, the actions of healthcare professionals may also involve how they speak to the caregivers. Hence, primary healthcare facilities should conduct a GMP service-delivery survey to enable users to provide feedback. In addition, it may be significant for clinic managers to address factors that should enhance the attitude of healthcare workers, which include in-service training; this is because their attitude affects the attitude of caregivers. A mixed-method study carried out in Ghana found an association between higher attendance and greater growth, which supports the need to enhance service delivery by strengthening primary healthcare institutions. Additionally, there is a need to raise awareness among caregivers about the value of growth monitoring and promotion for early infant development and encourage engagement through home visits [11]. This will give facilities the chance to reflect and act appropriately to enhance the delivery of these services. Among other things, the Sustainable Development Goals (SDGs) strive to minimize child malnutrition [31,32]. Adherence to the GMP sessions is helpful in the fight against child malnutrition, so it is crucial to make sure that primary healthcare centers have the resources necessary to provide dependable GMP services. In order for caregivers to consistently feel positive about this programme, it is crucial to address variables that may have an impact on their opinions or attitude. All children under the age of five must follow GMP to be healthy. However, the first two years of a child's life are particularly important [33]. These years have been described as a vital window of opportunity for ensuring that children are appropriately fed to support healthy growth [34]. Children may be more susceptible to malnutrition in the first two years if GMP is not applied properly [35].

Participants in the study also mentioned that they felt good about the programme and looked forward to attending sessions despite the difficulties of the health workers' behaviour and the irregular availability of GMP. Feeling good about this programme is greatest foundation and positive attitudes. It might encourage participation and adherence of GMP sessions. Feeling good about GMP may be related to participants' awareness of the significance of this programme and their unwavering love for their children. However, this attitude or feeling good about the GMP programme should be enhanced through the improved service delivery of this programme, including both monitoring and the promotion component of GMP. The study also suggests a programme to encourage the use of GMP services that is motivated by the caregivers' love for their grandchildren or children.

5. Conclusions

According to this study, caregivers are knowledgeable about the significance of GMP and consistently attend sessions because they value, care for, and love their children and grandchildren. The irregular provision of services and the behaviours of the healthcare professionals who are the implementers of this significant programme have an impact on caregivers' attitudes. It is important to investigate additional variables that might affect caregivers' attitudes. At the same time, primary healthcare facilities should conduct a GMP service-delivery survey to enable users to provide feedback, for the purpose of improving services. The study also suggests a programme to encourage use of GMP services that is motivated by caregivers' love for their grandchildren and/or children. Caregivers reported feeling good about this programme, which boosted their attitudes.

Implications: There is the need for an intervention to address factors that may have an impact on the attitude of caregivers, to achieve optimal use of the GMP programme.

Limitations: This study's findings demonstrated that the caregivers of children under the age of five were aware of the importance of the GMP programme, but they did not establish a relationship between knowledge and attitude. It also emphasized how caregivers' attitudes are impacted by healthcare personnel's attitudes, though it did not evaluate how severe an influence this has. Additionally, this research did not quantify or categorize the participants' attitudes. This study did not include those who were not enrolled for GMP at Sebayeng Clinic, which could have added a different perspective on attitudes.

Author Contributions: M.H.M. co-supervised the study and conceptualized the article. M.R. co-supervised the study and proofread the manuscript. N.M., S.M. and K.M. were project leaders of the overall study and were responsible for data collection and interpretation. All authors have read and agreed to the published version of the manuscript.

Funding: This research received no external funding.

Institutional Review Board Statement: The Turfloop Research Ethical Committee (TREC) approved the research and issued clearance certificate number TREC/238/2021:UG on 14 October 2021.

Informed Consent Statement: Informed consent was obtained from all subjects involved in the study.

Data Availability Statement: This article depends on the data gathered from the caregivers of children between 6–24 months in Sebayeng Clinic in Polokwane, Limpopo Province. The data generated or analysed during the current study are not openly accessible due to planned publication. However, they can be requested from the corresponding author by email: pitso85@gmail.com.

Acknowledgments: The caregivers of the children under the age of 5 years are appreciated and acknowledged for their voluntary participation.

Conflicts of Interest: The authors have no competing interest.

Disclaimer: The authors of this manuscript express their own views and not the position of the University of Limpopo.

References

1. Seidu, F.; Mogre, V.; Yidana, A.; Ziem, J.B. Utilization of growth monitoring and promotion is highest among children aged 0-11 months: A survey among mother-child pairs from rural northern Ghana. *BMC Public Health* **2021**, *21*, 910. [CrossRef] [PubMed]
2. United Nations Children's Fund. Malnutrition. 2020. Available online: https://data.unicef.org/topic/nutrition/malnutrition/ (accessed on 1 May 2022).
3. United Nations Children's Fund (UNICEF). World Health Organization and World Bank Group Joint Child Malnutrition Estimates. In *Levels and Trends in Child Malnutrition: Key Findings of the 2020 Edition*, 2020 ed.; UNICEF, Ed.; UNICEF: Geneva, Switzerland, 2020.
4. Keding, G. Nutrition transition in rural Tanzania and Kenya. *Hidden Hunger.* **2016**, *115*, 68–87.
5. Mkhize, M.; Sibanda, M. A Review of Selected Studies on the Factors Associated with the Nutrition Status of Children Under the Age of Five Years in South Africa. *Int. J. Environ. Res. Public Health* **2020**, *17*, 7973. [CrossRef] [PubMed]
6. Modjadji, P.; Madiba, P. Childhood undernutrition and its predictors in a rural health and demographic surveillance system site in South Africa. *Int. J. Environ. Res. Public Health* **2019**, *16*, 3021. [CrossRef] [PubMed]
7. Liu, Q.; Long, Q.; Garner, P. Growth monitoring and promotion (GMP) for children in low- and middle-income countries. *Cochrane Database Syst Rev.* **2017**, *1*, CD010102. [CrossRef]
8. Sulley, I.; Abizari, A.R.; Ali, Z.; Peprah, W.; Yakubu, H.G.; Forfoe, W.W.; Saaka, M. Growth Monitoring and Promotion practices among health workers may be suboptimal despite high knowledge scores. *BMC Health Sci. Res.* **2019**, *1*, 267. [CrossRef]
9. Mangasaryan, N.; Arabi, M.; Schultink, W. Revisiting the concept of growth monitoring and its possible role in community-based nutrition programs. *Food Nutr. Bull.* **2011**, *32*, 42–53. [CrossRef]
10. UNICEF (Ed.) *Revisiting Growth Monitoring and Its Evolution to Promoting Growth as a Strategic Program Approach: Building Consensus for Future Program Guidance*; UNCEF: Geneva, Switzerland, 2007.
11. Agbozo, F.; Colecraft, E.; Jahn, A.; TJBn, G. Understanding why child welfare clinic attendance and growth of children in the nutrition surveillance programme is below target: Lessons learnt from a mixed methods study in Ghana. *BMC Nurs.* **2018**, *17*, 1–14. [CrossRef]
12. Ashworth, A.; Shrimpton, R.; Jamil, K. Growth monitoring and promotion: Review of evidence of impact. *Matern. Child Health* **2008**, *4*, 86–117. [CrossRef]
13. Bilal, S.M.; Moser, A.; Blanco, R.; Spigt, M.; Dinant, G.J. Practices and challenges of growth monitoring and promotion in ethiopia: A qualitative study. *J. Health Popul. Nutr.* **2014**, *32*, 441–451.
14. Faber, J.S.; Al-Dhahir, I.; Reijnders, T.; Chavannes, N.H.; Evers, A.W.M.; Kraal, J.J.; van den Berg-Emons, H.J.G.; Visch, V.T. Attitudes Toward Health, Healthcare, and eHealth of People With a Low Socioeconomic Status: A Community-Based Participatory Approach. *Front. Digit. Health* **2021**, *3*, 690182. [CrossRef] [PubMed]
15. Vargas-Sánchez, A.; Plaza-Mejía, M.Á.; Porras-Bueno, N. Attitude. In *Encyclopedia of Tourism*; Jafari, J., Xiao, H., Eds.; Springer: Cham, Switzerland, 2016. [CrossRef]
16. Muchiri, J.W.; Gericke, G.J.; Rheeder, P. Effect of a nutrition education programme on clinical status and dietary behaviours of adults with type 2 diabetes in a resource-limited setting in South Africa: A randomised controlled trial. *Pub. Hea. Nutr.* **2016**, *19*, 142–155. [CrossRef] [PubMed]
17. Pampel, F.C.; Krueger, P.M.; Denney, J.T. Socioeconomic Disparities in Health Behaviors. *Annu. Rev. Sociol.* **2010**, *36*, 349–370. [CrossRef] [PubMed]
18. Skogen, J.C.; Bøe, T.; Finserås, T.R.; Sivertsen, B.; Hella, R.T.; Hjetland, G.J. Lower Subjective Socioeconomic Status Is Associated With Increased Risk of Reporting Negative Experiences on Social Media. Findings From the "LifeOnSoMe"-Study. *Front. Public Health* **2022**, *13*, 1749. [CrossRef]
19. Ramathuba, D.U.; Ndou, H. Ethical conflicts experienced by intensive care unit health professionals in a regional hospital, Limpopo province, South Africa. *Health SA Gesondheid* **2020**, *25*, 1–9. [CrossRef]
20. Creswell, J.W. *Qualitative Inquiry and Research Design: Choosing among Five Approaches*, 3rd ed.; Sage: Thousand Oaks, CA, USA, 2013.
21. Lee, M.; Kang, B.A.; You, M. Knowledge, attitudes, and practices (KAP) toward COVID-19, a cross-sectional study in South Korea. *BMC Public Health* **2021**, *21*, 295. [CrossRef]
22. Zhong, B.-L.; Luo, W.; Li, H.-M.; Zhang, Q.-Q.; Liu, X.-G.; Li, W.-T.; Li, Y. Knowledge, attitudes, and practices towards COVID-19 among Chinese residents during the rapid rise period of the COVID-19 outbreak: A quick online cross-sectional survey. *Int. J. Biol. Sci.* **2020**, *16*, 1745–1752. [CrossRef]
23. Charlton, K.E.; Kawana, B.M.; Hendricks, M.K. An assessment of the effectiveness of growth monitoring and promotion practices in the Lusaka district of Zambia. *Nutrition* **2009**, *25*, 1035–1046. [CrossRef]
24. Debuo, T.D.; Appiah, P.K.; Kweku, M.; Asalu, G.A.; Ahiabor, S.Y.; Takramah, W.K.; Duut, A.B. Caregivers Knowledge, Attitude and Practices on Child Growth Monitoring and Promotion Activities in Lawra District, Upper West Region of Ghana. *Sci. J. Public Health* **2017**, *5*, 20–30. [CrossRef]

25. Tekle, M.; Tariku, B.; Alagaw, A.; Zerihun, E.; Bekele, H.W. Exploring Reasons for Low Attendance of Mothers to Growth Monitoring and Promotion Program at Loka Abaya District, Southern Ethiopia: Exploratory Qualitative Study. *J. Nutr. Metab.* **2019**, *2019*, 3510649. [CrossRef]
26. Lewis, M.; McNaughton, S.A.; Rychetnik, L.; Chatfield, M.D.; Lee, A.J. Dietary Intake, Cost, and Affordability by Socioeconomic Group in Australia. *Int. J. Environ. Res. Public Health* **2021**, *18*, 13315. [CrossRef] [PubMed]
27. Oberoi, S.; Chaudhary, N.; Patnaik, S.; Singh, A. Understanding health seeking behavior. *J. Family Med. Prim. Care* **2016**, *5*, 463–464. [CrossRef] [PubMed]
28. Ogundele, O.A.; Ogundele, T.; Olajide, O.S.; Agunbiade, O.I. Effect of com- munity integrated management of childhood illness on mothers' healthcare-seeking behaviour and home management of childhood illness in Ile Ife, South-West Nigeria: A household survey. *S. A. J. Child Health* **2016**, *10*, 16–19. [CrossRef]
29. Sigdel, D.; Onta, M.; Bista, A.P. Factors affecting health seeking behaviors for common childhood illnesses among rural mothers in Chitwan. *Int. J. Health Sci. Res.* **2018**, *8*, 177–184.
30. Yaya, S.; Odusina, E.K.; Adjei, N.K. Health care seeking behaviour for children with acute childhood illnesses and its relating factors in sub-Saharan Africa: Evidence from 24 countries. *Trop. Med. Health* **2021**, *49*, 1–8. [CrossRef] [PubMed]
31. Sharma, D. Achieving sustainable development nutrition targets: The challenge for South Asia. *J. Glob. Health* **2020**, *10*, 010303. [CrossRef]
32. Grosso, G.; Mateo, A.; Rangelov, N.; Buzeti, T.; Birt, C.; on behalf of the Food and Nutrition Section of the European Public Health Association. Nutrition in the context of the Sustainable Development Goals. *Eur. J. Public Health* **2020**, *30*, i19–i23. [CrossRef]
33. Kebede, G.G.; Dawed, Y.A.; Seid, K.A. Child growth monitoring and promotion practice and associated factors among health care workers at public health facilities in south Wollo Zone, Northeast Ethiopia: A facility-based cross-sectional study. *BMC Nutr.* **2022**, *8*, 99. [CrossRef]
34. Nsiah-Asamoah, C.; Pereko, K.K.A.; Intiful, F.D. Nutritional counselling interactions between health workers and caregivers of children under two years: Observations at selected child welfare clinics in Ghana. *BMC Health Serv. Res.* **2019**, *19*, 817. [CrossRef]
35. Gyampoh, S.; Otoo, G.E.; Aryeetey, R.N.O. Child feeding knowledge and practices among women participating in growth monitoring and promotion in Accra, Ghana. *BMC Pregnancy Childbirth* **2014**, *14*, 180. [CrossRef]

children

MDPI

Article

HIV Status Disclosure to Adolescents Who Are Perinatally Infected in Rustenburg Sub District Northwest Province

Happy Maybe Maambiwa Khangale, Ndidzulafhi Selina Raliphaswa * and Azwidihwi Rose Tshililo

Department of Advanced Nursing Science, University of Venda, Private Bag X 5050, Thohoyandou 0950, South Africa
* Correspondence: ndidzulafhi.raliphaswa@univen.ac.za

Abstract: Acquired Immune Deficiency Syndrome (AIDS) is a viral disease caused by Human Immunodeficiency Virus (HIV) which affects the immune system of human body. This study sought to explore how adolescents with perinatal HIV infection learn about their status as well as investigate their preferences about the disclosure process. A qualitative exploratory, descriptive, and contextual research design was used to explore the disclosure of an HIV-positive status among adolescents on antiretroviral therapy. Nonprobability purposive sampling was used to select the healthcare facilities, and adolescents were chosen using convenience sampling. In-depth individual interviews were used to collect data from the participants until data saturation was reached. Collected data were analysed using Tesch's eight steps. The results of the study revealed that adolescents have been on ART (antiretroviral therapy) without the knowledge of their own status but taking ART. Delayed disclosure of an HIV-positive status to adolescents lead to adolescent not adhering to ART and wondering why they are on treatment while other adolescents are not. The study further revealed that parents and caregivers struggle to disclose an HIV-positive status of their children, leading them to lie about what the treatment is for, for example, that it is a treatment for cough.

Keywords: adherence; adolescents; antiretroviral therapy; disclosure; perinatally; parents; caregivers

1. Introduction

A growing percentage of HIV-infected children are surviving into adolescence because of the expansion of antiretroviral therapy (ART). According to the World Health Organization (WHO) and the joint United Nations programme on HIV/AIDS (UNAIDS), in 2020, there were 37.7 million [30.2–45.1 million] people living with HIV, including 36.0 million [28.9–43.2 million] adults and 1.7 million [1.2–2.2 million] children (0–14 years). In addition, 53% of all people living with HIV were women and girls. Of all people living with HIV, 84% [67–98%] knew their HIV status in 2020, and about 6.1 million [4.9–7.3 million] people did not know that they were living with HIV in 2020.

One of the elements linked to better adherence is the disclosure of HIV status. However, the prevalence of disclosure in children and adolescents varies depending on the environment and the age of the patients, ranging from 13 to 60% in children between the ages of 5 and 17 from Asia or Southern-Eastern Africa [1].

For adolescents, confidentiality may be jeopardized by disclosure because it may exacerbate mental and behavioural issues, familial disputes, or perceptions of societal stigma [2].

A recent study from Zimbabwe showed that learning about their HIV status is still one of the most difficult life events for adolescents living with HIV/AIDS [3]. Despite these facts, the World Health Organization's (WHO) guidelines for HIV status disclosure are limited to children under the age of 12 [4], even though many perinatally-infected children are not disclosed of their status until they become adolescents [5]. Healthcare workers and caregivers have minimally-tailored guidance on how to approach the issue of disclosure to these adolescents, other than that full disclosure is encouraged and should occur in "developmentally appropriate" stages [6]. The WHO recently developed new guidelines for HIV testing and counselling in adolescents [7], however, these guidelines do not address the issue of disclosure to adolescents, and only deal with the disclosure of adolescents' HIV statuses to others [8]. For disclosure to adolescents, the new guidelines simply defer to the above-mentioned existing guidelines for children under 12 [9].

The delayed disclosure of a child's HIV-positive status during the transition to adolescence affects their adherence to treatment, causing a high number of unsuppressed adolescents. Hence, we assessed and explored the knowledge of adolescents' own HIV-positive status.

2. Materials and Methods

2.1. Study Design

The study adopted a qualitative exploratory, descriptive, and contextual design to explore the disclosure of an HIV-positive status among perinatally-infected adolescents on ART. The design was chosen as the participants can freely narrate their experiences of being on ART treatment without the knowledge of what the treatment is about.

2.2. Data Collection

The study was conducted utilizing a qualitative research approach, which means that the qualitative data-gathering method was used to collect information from the participants. The study was conducted in one of the South African Provinces in the Northwest. A semi-structured interview was utilized as one of the qualitative data-collection techniques. The interviewer asked two questions followed by additional probing, and both close-ended and open-ended questions were included in the semi-structured interview in Appendix A.

Adolescents were recruited to participate in the study during their follow-up visit to the healthcare facilities in the Rustenburg sub-district of the Northwest Province. A pretest was conducted before the actual study began to detect and correct any flaws that might occur in the actual study. However, participants who were part of the pre-test were not included in the main study. The researcher scheduled appointments with the guardians of all participants under the age of 18 so that they could sign the consent form for their children to participate in the study. After receiving complete information regarding the research study (the benefits and risks), informed consent was also obtained from participants aged 18 and 19 years. This was completed after the adolescents on antiretroviral therapy between the ages of 15 and 19 years were informed about the study. Participants were also informed that participation was voluntary, and that they were able to withdraw from the study at any time.

The participants who consented to participate were asked to give their consent for a tape recorder to be used to record the interviews to capture all the information provided by the participants. The researchers purposively collected data from the sampled adolescents. Only 14 participants were included in the study as data was saturated at the 10th participant. However, interviews were continued until the 14th participant. This was completed to ensure that the data had indeed saturated, as no new information was provided. The research interview with the participants took place in their respective homes and in the research facilities. Each interview lasted for at least 30–45 min. All measures of trustworthiness were ensured including credibility, dependability, transferability, and conformability. Member check was also completed by going back to the participants with captured data so that they could confirm that the information that had been captured was the true reflection of what they said. All measures to reduce the spread of COVID-19 were adhered to throughout the study including hand washing, sanitizing, maintaining a social distance of 1.5 m, and masking.

2.3. Ethical Considerations

Approval to conduct the study was obtained from the University of Venda ethical clearance committee SHS/20/PDC/40/2110 and the approval date is 23 October 2020. The Department of Health, Northwest Province also gave approval to access the selected health facilities.

2.4. Data Analysis

In this study, individual verbatim comments from tape recordings were played and transcribed, and those delivered in Tswana were translated into English. Each transcription was examined separately before being discussed with the supervisors, as well as the final arrangement of the themes and sub-themes. Tesch's eight steps of the open coding method were used to analyse the data in this study. As stated by Lincoln and Guba (1985) and Krefting (1991), researchers used different techniques to ensure data trustworthiness, including credibility, dependability, conformability, and transferability. The researcher spent time with each adolescent to build trust. Purposeful probing was used for member checking throughout the interview. The researcher's findings were discussed with the participants. For dependability, experts were used to validate the methodology, which was further enhanced using an independent coder to ensure consistency. To ensure conformity, notes were retained in a secure location to allow for the creation of an acceptable trail and the determination of conclusions, interpretations, and suggestions if their sources could be tracked.

3. Results

3.1. Biographical Information of Participants

The study comprised 14 participants, all of whom were adolescents, who were willing to participate and met the criteria as indicated in Table 1 below. Of the 14 participants, 5 were less than 18 years old and 9 were 18 to 19 years old. The majority of participants (11) were on ART for more than 10 years whereas 3 were on ART for more than 2 years. Furthermore, 12 participants were still in secondary school whereas 2 were in postsecondary school.

Table 1. Biographical information of the participants.

Participant Number	Age	Gender	Number of Years on Treatment	Level of Education	Perinatal Acquired HIV
1	18	F	16+	Grade 12	Perinatal
2	17	M	15+	Grade 12	Perinatal
3	18	F	2+	Grade 11	Perinatal
4	19	M	16+	Postsecondary	Perinatal
5	18	F	17+	Grade 12	Perinatal
6	18	M	10+	Grade 12	Perinatal
7	17	F	15+	Grade 11	Perinatal
8	18	M	17+	Grade 11	Perinatal
9	17	F	16+	Grade 11	Perinatal
10	17	F	15+	Grade 10	Perinatal
11	19	F	17+	Grade 12	Perinatal
12	17	F	16+	Grade 10	Perinatal
13	19	F	2	Postsecondary	Perinatal
14	18	M	12	Grade 12	Perinatal

The study findings revealed one major theme accompanied by five sub-themes as indicated in Table 2 above. The major theme is knowledge related to HIV+ status and related factors, lack of versus the existence of knowledge related to the importance of adherence to treatment, lack of versus the existence of knowledge related to the initial diagnosis, side effects, lack of knowledge related to the HIV+ status and how it was transmitted, and disclosure of HIV-positive status viewed as important.

Table 2. Themes and sub-themes that emerged from the study findings.

1. Knowledge related to HIV+ status and related factors	1.1	Lack of versus the existence of knowledge related to the disease condition from childhood to date explained
	1.2	Lack of versus the existence of knowledge related to the importance of adherence to treatment and its future implications described
	1.3	Lack of versus the existence of knowledge related to the initial diagnosis, side effects, signs, and symptoms explained
	1.4	Lack of knowledge related to the HIV+ status and how it was transmitted (parents blamed)
	1.5	Disclosure of HIV-positive status viewed as important

3.2. *Theme 1: Knowledge Related to HIV+ Status and Related Factors*

Theme 1 is divided into five sub-themes which describe the knowledge related to HIV+ status and related factors.

3.2.1. Sub-Theme 1.1: Lack of Versus the Existence of Knowledge Related to the Disease Condition from Childhood to Date Explained

Participants in this study revealed that they did not have knowledge related to their HIV status and other factors related to it. This was supported by participants who expressed themselves as follows:

> Participant 1 said, "(Looking down angrily) I found out a few months ago when I went to get my prescription with my mother, and she took me to the counselling room with her, where I was told everything, I never thought I might be HIV positive as my mother didn't tell me the truth about the medication that I have been using every day. Telling me the truth while we were at home could have at least been something to me (crying)".

One participant only learned about her own perinatally-acquired HIV-positive status after being sick and visiting a local clinic, where her status was revealed to her.

> Participant 7 said, "I was sick and came to the clinic and tested positive for HIV and when I arrive home my mother was not surprised because she knew I was born with it. I couldn't even look at her as I was disappointed in her for not taking care of me, for not disclosing to me about my status and make sure I take treatment. I had many questions as to why she could let me be sick while she knows the truth behind my sickness. Learning about ones HIV positive status and find only found out that you were born with it felt like a betrayal to me by my mother".

3.2.2. Sub-Theme 1.2: Lack of Versus the Existence of Knowledge Related to the Importance of Adherence to Treatment and Its Future Implications Described

Participants specified that they could see the benefits of adhering to ART, as they were healthy and stronger with treatment. Participants responded in the following ways:

> Participant 12 said, "Yes, I see the benefits because when I'm on treatment, people don't realize I'm sick because my body is healthy. Taking treatment have always made my blood results to me good when I am checked. The nurse's health education on treatment importance played a major role for me to know that I

> shouldn't skip treatment doses. Treatment prevented me from being sick by other diseases".

> Participant 10 said, "Yes, I see the benefits because treatment has made me stronger from birth. The treatment is the one which have enabled me to transition from childhood to adulthood. It made me to still look like other adolescents although I am HIV positive. The treatment increases lifespan".

The participant further indicated that although there are benefits of treatment, he also experiences side effects which he does not regard as a benefit. This is how he expressed himself:

> "Yes, there are benefits; for example, you don't get sick as much when you're on treatment (looking down); but there are times where there are no benefits because I sleep and don't finish my schoolwork. Taking treatment early results in me falling early that's why I hardly complete schoolwork. It becomes tough because my mother does not allow me to change times as we were told at the clinic that treatment is taken only in one time so that it can work for me".

> Participant 5 said, "I see the benefits, but I have my doubts because I am receiving treatment and instead of studying, I am sleeping, therefore I am not benefiting. Sleeping result in me not performing well at school which is not good for my future".

> Participant 8 said, "Yes, there are additional benefits to receiving treatment, such as being protected from becoming ill and making it difficult for others to notice that you are HIV positive. I was told by the nurse that when I don't take treatment everyone will know that I am sick. My treatment makes my blood to be strong. I have faith in life because of my treatment".

The benefits of recovering from opportunistic infections by being on treatment was also indicated by the participant.

> Participant 3 said, "Yes, I see more benefits because I was sick and ended up with TB when I wasn't taking treatment, and I've seen the benefits of taking treatment after starting treatment. The time I was sick not taking treatment I had given up in life with no hope, but the treatment brought life and hope to me as I have been well since taking it. I have indeed experienced it that ART increases one's lifespan, I wouldn't be living today if it was not for the treatment. It made my cd4 cells to be strong".

3.2.3. Sub-Them 1.3: Lack of Versus the Existence of Knowledge Related to the Initial Diagnosis, Side Effects, Together with the Signs and Symptoms Explained

The study revealed that adolescents lack knowledge related to the initial diagnosis of HIV. The side effects together with the signs and symptoms after the initial diagnosis were explained by different participants who emphasized the difficulty of engaging in their daily routines after taking treatment. The following quotes were made:

> Participant 4 said, "I was struggling to cope with the side effects after taking treatment since I couldn't do anything but sleep, especially schoolwork, which I couldn't do after taking treatment, but now that the school day has changed, I only have trouble waking up and feeling sleepy in the mornings".

> Participant 3 said, "I've also failed in school since I've been unwell and missed school days, this was because of the delay of taking treatment which made me to have opportunistic infections that required me to be hospitalized. When I started treatment, I had no energy to study after taking treatment all I want was to sleep. I experienced the feeling of nausea and dizziness the first week of starting treatment".

> Participant 13 said, "As for me, I've noticed that being positive might make you feel different from other children, particularly in the first months following

> disclosure. You always feel like you no longer worthy to be amongst others, what if they find out? Will they accept my condition. The fear of losing my loved ones in the community due to my status took over in the first month of disclosure. The dizziness I have after treatment forces me to go to bed early, sometimes I don't manage to do all my schoolwork".

In this study, participants had difficulties coping with the side effects during the first month of diagnosis. Therefore, the researchers observed that support for adolescents during their first month of treatment is important for them to endure the side effects and regain hope after being diagnosed with a lifelong condition.

3.2.4. Sub-Theme 1.4: Lack of Knowledge Related to the HIV+ Status and How It Was Transmitted (Parents Blamed)

The participants in the study indicated that they lack knowledge about their HIV+ status. Adolescents took treatment without the knowledge of their status, which led to them to put more blame on their parents after disclosure of their status and the way in which HIV was transmitted to them. Participants who were infected through mother-to-child transmission believed that their mothers would have taken treatment to prevent mother-to-child transmission.

> Participant 6 said, "Yes, she did after a bit of coercion. My mother had been on ART for years and informed me when she was pregnant that she wasn't taking her treatment properly, which resulted in her giving birth to a positive baby, which was me, and she never made me to take treatment; instead, she remained silent and waited for me to find out on my own. It was never easy to me to accept because she could have prevented me from being infected by adhering to treatment while she was pregnant. I felt like my mother betrayed me and didn't love me enough she could have at least taken treatment to give birth to me without HIV positive (angry)".

While some participants explained how HIV was transmitted to them, other participant had the questions regarding why they were born negative, but tested positive for HIV after a few months of breastfeeding.

> Participant 10 said, "Yes, she did inform me that I was born HIV negative, but that I tested positive for HIV/AIDS within six months of breastfeeding. It made it difficult for me to comprehend the process of becoming positive while breastfeeding. I had many questions as how come and my mother told me that her blood was not well which made me to be infected. My mother should have at least not breastfed. Her blood was not good because she didn't adhere to treatment".

> Participant 9 said, "My mother told me when I was 15 years old that she was diagnosed with HIV as she was about to give birth in her last month, which caused her to give birth to an HIV-positive child, of which I am the child. She said she was not using condom as she was pregnant which made her to be infected with HIV. They tried to give her treatment after being diagnosed but it was late to save me from getting HIV".

Adolescents with perinatal HIV were dissatisfied with how they contracted the virus and blamed their parents for not doing more to prevent them from contracting it.

3.2.5. Sub-Theme 1.5: Disclosure of HIV-Positive Status Viewed as Important

Participants in the study thought it was crucial to disclose their HIV-positive status because it affects adherence to their treatment. Knowing why they need to take treatment motivates them more than not knowing. Participants in the study shared their experiences before and after disclosure, emphasizing the importance of disclosure. This is supported by the following quotes from the participants:

> Participant 2 said, "Yes, it was difficult because I had no idea it could be HIV treatment, and as a child, I had to be furious that I was not told and that I was constantly forced to take treatment that I had no idea what it was for. I was angry towards my parents for not telling me what the treatment was for. I felt like I was betrayed by not being told early that I am HIV positive. To be told you are HIV-positive at my age it was really hurting and not easy to accept as I had many thoughts as to what my friends will say if they find out about my status (looking down). I thought I wouldn't make it face the world".

The process of disclosure was delayed by parents and caregivers as they had to lie about their children's status and not tell them the truth about the treatment. Poor disclosure to adolescents leads to poor adherence to treatment.

> Participant 5 said, "I couldn't take my treatment before disclosure because I didn't understand why, and the treatment I was taking was so hot I couldn't stand swallowing it, so I didn't take it every day. Some days I pretended to take it when I didn't, and after disclosure, I was able to take treatment knowing exactly what it was for. It was not easy to take treatment before disclosure because other children of my age were living freely. The disclosure helped me to adhere to treatment as I knew the danger of poor adherence. Although I was angry at my parents I for lying to me I continued taking treatment better than before".

Another participant, despite being diagnosed with HIV during adolescence, still disclosed their status to family and received support rather than discrimination and stigma.

> Participant 14 said, "I had support from my sister, who was able to sit with me and show me, love, by correcting me and telling me that no matter what had occurred, I should not lose hope and that there is still life even if I am HIV positive. I was afraid to disclose knowing how careless I was not protecting myself. My sister showed me love and support which gave me hope to adhere to treatment".

4. Discussion

Participants in the study revealed a lack of understanding of their disease condition, as they only learned about it after visiting a healthcare facility because they were sick, when it was then revealed to them that they were HIV+. Moreover, other participants had been taking treatment without realizing it was ART, which was only revealed to them when they went for collection.

The study findings are supported by the study conducted by [10] which revealed that most of the individuals had been ill for a long time and had been hospitalized several times before receiving their ART treatment. A majority of the adolescents were still at school, therefore, their school activities were negatively affected because of their illness. Consequently, some of the adolescents had to repeat classes because they were unable to attend classes on a regular basis. Hence, studying was difficult for them due to their illness.

The study further revealed that adolescents had no knowledge regarding their HIV status until they developed opportunistic infection symptoms which led them to visit the clinics where they were diagnosed HIV+. However, knowing their HIV status helped them to be more positive about taking the treatment, even though it was not easy, since they underwent denial before accepting the situation. Knowing their HIV status allowed the participants to embrace the treatment's importance in their lives, as to them, it appears that it is their reason for living—they have put their faith in receiving treatment because it keeps them healthy.

These findings are consistent with those of a study conducted by [10] which highlighted that an individual's experiences determine adherence and non-adherence. Furthermore, the study revealed that an adolescent's parents, caregivers, and healthcare providers influenced their treatment adherence by instilling in them a fear of dying. Therefore, the researchers observed from this study that adherence to treatment was viewed as important by most participants who believed that, if it was not for treatment, they would not be living.

Contrary to the above finding, only one participant in the current study was perplexed by the importance of treatment, viewing it as significant when it did not allow him to complete his schoolwork, which was supposed to help him construct a future. Therefore, based on the above information, it is critical to emphasize the importance of health education on side effects to avoid noncompliance.

Furthermore, this study revealed that when adolescents are aware of their HIV status, they are better able to comprehend the importance of hospital visits and drug regimens, allowing them to make meaning out of their lives [11].

According to [12], adolescents living with HIV infection in Zambia reported going through changes after learning about their HIV-positive status, including enduring the obstacles associated with being infected with HIV/AIDS. After being diagnosed with a life-threatening illness, these children were able to survive. On the other hand, some of the children recounted feelings of despair after learning of their HIV-positive diagnosis, which came after a long time of illness. Consequently, the disclosure was both alarming and disempowering for these children, which required the need for counselling support. Regardless, caregivers also extended the revelation of their HIV-positive status which was influenced by mistrust, worries of stigma and discrimination, and secrecy [13]. Similarly, the findings of this study concurred with the findings of the study conducted by [14], who affirmed that participants displayed some anger and an attitude of blame toward their biological parents for infecting them with the virus. This is also consistent with the findings of the study by [15], who reported that children born with HIV who survive into adolescence are at risk of developing psychological disturbances because of long-term HIV and its related stressors, together with the long-term effects of medications.

For a variety of reasons, it is critical to inform children with HIV about their HIV status. The fact that they have been diagnosed with HIV is significant, and it is now part of their complete HIV care. According to some research, children who are told about their HIV infection are more likely to stick to their antiretroviral therapy (ART) [16], and [17] children who are disclosed of their status are four times more likely than nondisclosed children to adhere to ART. As a result, in both developing and developed countries, nondisclosure to children and adolescents with HIV has been highlighted as a barrier to adherence to ART [18]. Norms, taboos, and beliefs about talking about sex, a lack of family guidance by HCWs to disclose the HIV status to their children, and lack of family guidance by HCWs must all play a role in breaking the taboos and norms of caregivers of talking about sexuality and disclosure to their children to facilitate adherence to ART. To make an informed decision to take medication, adolescents must first comprehend the consequences of non-adherence on their health and well-being. The findings of this study are similar to those of [16], which discovered that disclosing HIV improves knowledge and understanding of the condition while also lowering risky behaviours. Unprotected sexual intercourse and unintentionally infecting a partner are two risky activities that adolescents engage in. When people are informed of their HIV status, they can take steps to protect their health and reduce the risk of infecting others [19].

5. Conclusions

This study revealed an essential issue regarding HIV status disclosure to adolescents. It is important for communities to be educated on the importance of continuous HIV testing, knowledge, and understanding of HIV and ART. This should include the signs and symptoms of HIV/AIDS together with opportunistic infections. The information aimed to assist all individuals in limiting their shock and confusion after being diagnosed with HIV. Parents also need to know the importance of disclosing their HIV status to their adolescents, before they find out the information on their own, to prevent blame and anger.

Therefore, each circumstance should be examined separately to make unique disclosure recommendations. Although many other factors were not included in this study and the disclosure experiences emphasized in this study were not substantial predictors of HIV status, disclosure to adolescents who are perinatally infected remains crucial. In the future, researchers should consider utilizing other research techniques to draw more thorough conclusions on the disclosure of HIV status to adolescents who are perinatally infected.

6. Recommendations

The medical and social support staff should conduct continuous counselling to promote early disclosure to parents raising children with perinatal-acquired HIV. Campaigns should also be conducted in communities to educate the public about HIV, reducing the stigma directed at people on ART. Future research should investigate the experiences of parents and guardians while disclosing the HIV-positive status of their children and the challenges they are facing. The department should strengthen support to parents during the process of disclosure of an HIV-positive status of adolescents.

Author Contributions: Conceptualization, H.M.M.K. and N.S.R.; Methodology, A.R.T.; Software, A.R.T.; Validation, H.M.M.K., N.S.R. and A.R.T.; Formal analysis, A.R.T.; Investigation, H.M.M.K.; Writing—original draft preparation, H.M.M.K.; Writing—review and editing, A.R.T.; Visualization, A.R.T.; Supervision, N.S.R. All authors have read and agreed to the published version of the manuscript.

Funding: This research received no external funding.

Institutional Review Board Statement: Approval was obtained from the ethical clearance committee at the University of Venda and from the Provincial Department of Health, Northwest Province.

Informed Consent Statement: Informed and assent consent was obtained from all subjects involved in the study.

Data Availability Statement: Data is available on request.

Acknowledgments: All participants are thanked for their time and willingness to participate.

Conflicts of Interest: Authors declare that there are no conflict of interests.

Disclaimer: Views and opinions remain the responsibility of the authors and have not been submitted to any journal for publication.

Appendix A

Semi-structured interview:

1. When did you start taking ART?
2. What are your experiences while on ART?

Appendix B

CONSENT:

Statement of Agreement to Participate in the Research Study:

- I, ...hereby confirm that I have been informed by the researcher, Khangale Maambiwa, about the nature, conduct, benefits, and risks of this study—Research Ethics Clearance Number:
- I have also received, read, and understood the above written information (Participant Letter of Information) regarding the study.
- I am aware that the results of the study, including personal details regarding my sex, age, date of birth, initials, and diagnosis will be anonymously processed into a study report.
- In view of the requirements of research, I agree that the data collected during this study can be processed in a computerized system by the researcher.
- I may, at any stage, without prejudice, withdraw my consent and participation in the study.
- I have had sufficient opportunity to ask questions and (of my own free will) declare myself prepared to participate in the study.

- I understand that significant new findings developed during the course of this research which may relate to my participation will be made available to me.

Full Name of Participant Date Time Signature

I, ...

Khangale Maambiwa herewith confirm that the above participant has been fully informed about the nature, conduct and risks of the above study.

Full Name of Researcher

.... Date............................ Signature.........................

Full Name of Witness (If applicable)

.... Date .... Signature.............................

Full Name of Legal Guardian (If applicable)

.... Date........................ Signature..................................

Please note the following:

Research details must be provided in a clear, simple, and culturally-appropriate manner and prospective participants should be helped to arrive at an informed decision by use of appropriate language (grade 10 level-use Flesch Reading Ease Scores on Microsoft Word), selecting of a nonthreatening environment for interaction and the availability of peer counselling (Department of Health, 2004)

If the potential participant is unable to read/illiterate, then a right thumb print is required and an impartial witness, who is literate and knows the participant, e.g., parent, sibling, friend, pastor, etc., should verify in writing, duly signed, that informed verbal consent was obtained (Department of Health, 2004).

If anyone makes a mistake completing this document, e.g., a wrong date or spelling mistake, a new document has to be completed. The incomplete original document has to be kept in the participant's file and not thrown away, and copies thereof must be issued to the participant.

References

1. García-Boyano, M.; Chávez-Solórzano, N.; Layana-Coronel, M.; Soffe-Pazmiño, J.; Sarcos-Lindao, H.; Solís-Montiel, D.; Miño-León, G. Determinants of Disclosure, Adherence and Viral Suppression in Children and Adolescents Living With HIV in Ecuador: A Cross-Sectional Study. *Pediatric Infect. Dis. J.* **2022**, *41*, e133–e138. [CrossRef] [PubMed]
2. Rich, C.; Mavhu, W.; France, N.F.; Munatsi, V.; Byrne, E.; Willis, N.; Nolan, A. Exploring the beliefs, experiences and impacts of HIV-related self-stigma amongst adolescents and young adults living with HIV in Harare, Zimbabwe: A qualitative study. *PLoS ONE* **2022**, *17*, e0268498. [CrossRef] [PubMed]
3. Haberer, J.E.; Cook, A.; Walker, A.S.; Ngambi, M.; Ferrier, A.; Mulenga, A.; Kityo, C.; Thomason, M.; Kabamba, D.; Chintu, C.; et al. Excellent adherence to antiretrovirals in HIV+ Zambian children is compromised by disrupted routine, HIV nondisclosure, and paradoxical income effects. *PLoS ONE* **2011**, *6*, e18505. [CrossRef] [PubMed]
4. Bikaako-Kajura, W.; Luyirika, E.; Purcell, D.W.; Downing, J.; Kaharuza, F.; Mermin, J.; Malamba, S.; Bunnell, R. Disclosure of HIV status and adherence to daily drug regimens among HIV-infected children in Uganda. *AIDS Behav.* **2006**, *10*, S85–S93. [CrossRef] [PubMed]
5. Menon, A.; Glazebrook, C.; Campain, N.; Ngoma, M. Mental health and disclosure of HIV status in Zambian adolescents with HIV infection: Implications for peer-support programs. *J. Acquir. Immune. Defic. Syndr.* **2007**, *46*, 349–354. [CrossRef] [PubMed]
6. Gray, G.E. Adolescent HIV–cause for concern in Southern Africa. *PLoS Med.* **2010**, *7*, e1000227. [CrossRef] [PubMed]
7. Rujumba, J.; Mbasaalaki-Mwaka, C.L.; Ndeezi, G. Challenges faced by health workers in providing counselling services to HIV-positive children in Uganda: A descriptive study. *J. Int. AIDS Soc.* **2010**, *13*, 9. [CrossRef] [PubMed]
8. De Baets, A.J.; Sifovo, S.; Parsons, R.; Pazvakavambwa, I.E. HIV disclosure and discussions about grief with Shona children: A comparison between health care workers and community members in Eastern Zimbabwe (English). *Soc. Sci. Med.* **2008**, *66*, 479–491. [CrossRef] [PubMed]
9. Creswell, J.W. *Qualitative Inquiry and Research Design: Choosing among Five Approaches*; Sage: Thousand Oaks, CA, USA, 2014.
10. Ramarumo, T. The Psycho-social Experiences of Children Living with Vertically Transmitted HIV/AIDS at Messina Hospital, South Africa. Doctoral Dissertation, 2018.
11. Midtbo, V.; Shirima, V.; Skovdal, M.; Daniel, M. How disclosure and antiretroviral therapy help HIV- infected adolescent in sub-Saharan Africa cope with stigma. *Afr. J. AIDS Res.* **2016**, *11*, 261–271. [CrossRef] [PubMed]
12. Mburu, G.; Hodgson, I.; Kalibala, S.; Haamujompa, C.; Cataldo, F.; Lowenthal, E.D.; Ross, D. Adolescent HIV disclosure in Zambia: Barriers, facilitators and outcomes. *J. Int. AIDS Soc.* **2014**, *17*, 18866. [CrossRef] [PubMed]
13. Harris, J.E. Taking disability public. *U. Pa. L. Rev.* **2020**, *169*, 1681.

14. Hayfron-Benjamin, A.; Obiri-Yeboah, D.; Ayisi-Addo, S.; Siakwa, P.M.; Mupepi, S. HIV diagnosis disclosure to infected children and adolescents; challenges of family caregivers in the Central Region of Ghana. *BMC Paediatr.* **2018**, *18*, 1–11. [CrossRef] [PubMed]
15. Malungani, N.F.; Rasweswe, M.M.; Peu, M.D. "I'm angry with my mother": Lived psychosocial experiences of adolescents with peri-natal-acquired HIV in Limpopo, South Africa. *J. Psychol. Afr.* **2021**, *31*, 627–631. [CrossRef]
16. Madiba, S. Disclosing HIV to Infected Children in South Africa in the Era of HAART: A Grounded Theory Study on the Process, Reasons and Outcomes of Disclosure. *World J. AIDS* **2016**, *2*, 319–329. [CrossRef]
17. Jemmott, J.B., III; Zhang, J.; Croom, M.; Icard, L.D.; Rutledge, S.E.; O'Leary, A. Barriers and facilitators to engaging African American men who have sex with men in the HIV care continuum: A theory-based qualitative study. *J. Assoc. Nurses AIDS Care* **2019**, *30*, 352–361. [CrossRef] [PubMed]
18. Arage, G.; Tessema, G.A.; Kassa, H. Adherence to antiretroviral therapy and its associated factors among children at South Wollo Zone Hospitals, Northeast Ethiopia: A cross-sectional study. *BMC Public Health* **2014**, *14*, 1–7. [CrossRef] [PubMed]
19. Biressaw, S.; Abegaz, W.E.; Abebe, M.; Taye, W.A.; Belay, M. Adherence to Antiretroviral Therapy and associated factors among HIV infected children in Ethiopia: Unannounced home-based pill count versus caregivers' report. *BMC Paediatr.* **2015**, *13*, 1–9. [CrossRef] [PubMed]

children

MDPI

Article

Perceptions of Caregivers Regarding Malnutrition in Children under Five in Rural Areas, South Africa †

Takalani. E Thabathi, Mary Maluleke *, Ndidzulafhi. S Raliphaswa and Thingahangwi. C Masutha

Department of Advanced Nursing Science, University of Venda, Private Bag X 5050, Thohoyandou 0950, South Africa

* Correspondence: marymaluleke@univen.ac.za

† This is part of Masters degree research work for Thabathi Takalani Eldah in University of Venda, Thohoyandou, Limpopo Province, South Africa.

Abstract: Children under five depend on the caregivers to provide them with nutritious food to combat malnutrition. Several studies have been conducted about malnutrition in children, yet limited attention has been paid to the caregivers. Thus, the study investigated the perceptions of caregivers with regard to non-nutritious foods in rural areas in South Africa. This study explored caregiver's perception regarding non-nutritious food in children below five. This was done in primary health care clinics of Tshilwavhusiku local areas of Makhado Municipality in Vhembe District, South Africa. A qualitative approach was adopted. Nine caregivers were sampled randomly. In-depth individual interviews were conducted, and Tesch's analytical approach was adopted to analyze the data measures to ensure trustworthiness and ethical considerations were adhered to throughout the study. The study revealed that caregivers were lacking knowledge regarding nutritious food to be given to their children and signs of malnutrition were also not known. It is imperative to ensure the improvement of caregivers' knowledge regarding nutritious food and children's nutritional status in order to reduce the malnutrition rate.

Keywords: caregivers; children under-five; malnutrition; perception

1. Introduction

According to the Children's Act 2005 [1], children under five are defined as people between 0 and 59 months. The Bill of Rights in the South African Constitution Section 28(1) (c) guarantees every under-five child the right to basic nutrition [2]. Nutrition plays an essential role in children's growth, development, health, psychosocial functioning, and productivity [3,4]. Good nutrition for children includes the right amounts of vitamins, minerals, carbohydrates, protein, and fats. Breast milk is recommended as nutritious food in children below six months of age as it contains carbohydrates, protein, and fats [5]. The complementary foods for children between six and 59 months should contain different types of food groups, i.e., starchy foods, protein, vegetables, fats, and oils, and they should drink 6–8 glasses of clean, safe water [6].

All children under five depend on caregivers to have sufficient nutritious food [7]. Caregivers are responsible for providing under-five children with nutritious food daily since under-five children cannot do that themselves. This is in line with children's act 38 of 2005, which defines caregivers as a person 16 years old and above, besides a parent or guardian, who can take care of the under-five children. The South African government supports children in need with the Child Support Grant (CSG), which is meant to buy nutritious food [8].

However, malnutrition in children under five is still alarming. Globally, among children under the age of five, 157 million are stunted, 41 million are overweight, and 50 million are wasted [9]. Africa also has a high prevalence of stunting, wasting, and being overweight. In 2019, two out of five stunted children and one quarter of the world's wasted and overweight children under the age of five years lived in Africa [9]. According to the South Africa Demographic and Health Survey (SADHS) of 2016, wasting was found to be 2.5%, underweight at 6%, and stunting was at 27% amongst children under the age of five years [10]. According to DHIS (District Health Information System), 24 of 111 (21, 6%) and 24 of 141 (17%) children in the Kutama Clinic of the Vhembe District were reported to experience malnutrition between 2014 to 2016 and 2018 to 2019 respectively.

Several studies have been conducted about malnutrition in children under five years. Cumber et al. [3] conducted a study about poor complementary feeding practices among young children in Cameroon. Murad et al. [11] investigated the understanding of factors affecting breastfeeding practices in one city in the Kingdom of Saudi Arabia. Mohseni et al. [12] focused on factors associated with nutrition among under-five-year-old children in Iran. Omotoye et al. [13] focused on Infant and young child- feeding practices in two local government areas in southwest Nigeria. Mushaphi et al. [14] conducted a study about recommendations for infant feeding policy and programs in Dzimauli region, South Africa. Motadi et al. [15] focused on breastfeeding knowledge and practices among mothers of children younger than two years from a rural area in the Limpopo Province, South Africa.

Malnutrition in Makhado municipality was a great concern. It occurred in the same child or same household in 13 out of 24 children in 2014 to 2016. This prompted the researcher to investigate the perceptions of caregivers regarding malnutrition in children under the age of five years in rural areas in South Africa.

2. Materials and Methods

2.1. Research Design

This paper adopted a qualitative approach. An explorative, descriptive, and contextual design was also adopted. The approach was suitable for this paper, since the caregivers explored their perceptions regarding malnutrition which could not be captured in numbers, and the participants narrated their lived experiences at their home which is their contextual setting.

2.2. Study Setting

The study was conducted in Primary Health Care clinics of Tshilwavhusiku local area of the Makhado Municipality in Vhembe District, Limpopo Province. Out of 115 fixed primary health care clinics in the Vhembe district, the study focused on only three clinics with a high rate of child malnutrition out of 6 clinics and one health center.

2.3. Study Population

The population of this study included caregivers of children who are malnourished and are under the age of five years.

2.4. Sampling Method

Sampling techniques were purposive and only caregivers with malnourished children were selected as they will be able to give full and rich information needed by the researcher. Three top clinics (Kutama, Tshilwavhusiku, and Madombidzha) out of seven in the Tshilwavhusiku local area with a high rate of malnutrition in children under five years of age were sampled purposively. The DHIS (district health information system) of Vhembe was the source of the top three clinics. Caregivers was selected purposively. From the list of all children with malnutrition in each clinic, Venda-speaking participants only with a full physical address and contact details were selected.

2.5. Sample Size

The researcher expected to sample 5 caregivers per clinic for a total of 15 for 3 selected clinics. However, data saturated at the 9th participant and the researcher continued until the 15th to ensure that there was no new information coming out. Their information was not included as it was a repetition of what had been already said by the first 9 participants.

2.6. Data Collection

Trustworthiness was ensured by applying the framework of [16]. The participants were visited by the researcher prior to the interviews to build rapport and trust. Appointments were made with the participants telephonically prior to the interviews by the researchers to tell them about the study. Then, caregivers' homes were visited to give them full information about the study, and to give them the opportunity to talk about their perceptions regarding malnutrition in children under five, as they are people with relevant information, which could assist the researchers to achieve the aim and objective of the study. The caregivers consented verbally to be part of the interview. Date and time that was convenient to them for the interview was scheduled.

Data were collected through in-depth individual interviews at the participant's homes at times favorable to caregivers. The open-ended interview was conducted in Tshivenda, guided by one central question: *"Can you kindly share with me what you think may be contributing to malnutrition in the child?"* Unstructured interviews are usually qualitative in nature, and can be very helpful for social science or humanities research focusing on personal experiences. Furthermore, participants are able to express their experiences and feelings freely. Each interview lasted for approximately 30–45 min. Effective communication skills were used: active listening to encourage the participants to speak more, reflection for participants to elaborate more, and clarification on statements not understood and this helped the participants to be at ease and give more information until no more new information was coming out from the interviews researcher to ensure credibility. A tape recorder was used throughout the interview as permitted by the participants to capture all information given to participants. Furthermore, at the end of each interview, the tape recorder was played back for the participants to verify the recorded information and to identify gaps, as a way to ensure member check. The purpose of the member check was to get rid of researcher bias when interpreting the results. The researcher withheld from influencing the participant's findings to ensure conformability. Research methods were fully described to ensure transferability. Transferability was obtained through a dense description of research methods. The research findings were also discussed with the participants. Data were transcribed verbatim.

2.7. Data Analysis

Data were analyzed by a team of researchers using Tesch's approach by Creswell [17]. The voice recorder was played and recorded information transcribed verbatim. The researchers read through the transcribed data several times to get the meaning of the responses from caregivers. Then, they arranged the data into subcategories and categories, labelled them using the actual words and language of the caregivers, and came up with the theme, which appears to be the major findings of the study.

The relation between the subcategories and categories was cross-checked to see if they relate to what caregivers said during interviews, referring to the recorder. The findings of the study were then interpreted; explaining what has been learned about caregivers' perceptions regarding malnutrition in children under the age of five. The interpretation of the findings was grounded by the researcher's understanding regarding malnutrition, with the integration of various kinds of literature regarding malnutrition.

2.8. Ethical Considerations

Prior to starting the research process, permission to conduct the study was obtained from the University of Venda Research Ethics Committee (SHS/18/PDC/05/1505), The Limpopo Department of Health Research Ethics Committee, as well as the Department of Health, Vhembe District. The researcher provided the participants with information regarding the study to have an idea of what consenting for. To ensure no harm, the researcher asked questions that are unoffending and adopted a non-judgmental attitude toward the participants. Codes were used instead of participants' actual names, to protect their identity. Information provided by the participants was kept away from people who do not form part of the study.

3. Results

The study involved a total of 9 caregivers, of which four were mothers and five were grandmothers. This is illustrated in Table 1.

Table 1. Age of Caregivers.

Age of Caregivers	Number of Caregivers	Relationship to the Child
20–29 years	02	Mothers
30–39 years	01	Mother
40–49 years	01	Mother
50–59 years	02	Grandmothers
60–69 years	03	Grandmothers

Most of the participants were grandmothers, more than half at the rate of 55.5%, and less than half (44.5%) were mothers.

A majority of caregivers were not employed and did not go to school at all. Some depended on social grants to feed. Those that managed to go to school only reach grade 12, as depicted in Table 2 above.

Table 2. Age, employment status and educational level of caregivers.

Age of Caregivers		Educational Level
20 Years	Unemployed	Grade 12
21 Years	Unemployed	Grade 8
33 Years	Working at the farm	Grade 11
49 Years	Self employed	Grade 2
58 Years	Unemployed	Not educated
59 years	Unemployed	Not educated
61 Years	Pensioner	Not educated
66 Years	Pensioner	Not educated
66 Years	Pensioner	Not educated

Two themes emerged from data analysis, which include a lack of knowledge regarding nutritious food and a knowledge deficit regarding signs of malnutrition.

3.1. *Theme 1: Lack of Knowledge Regarding Types of Nutritious Food*

Findings revealed that participants lacked knowledge regarding nutritious food. Two subthemes emerged from the theme (lack of knowledge regarding types of nutritious food) namely: breastfeeding and non-nutritious food.

3.1.1. Sub-Theme 1.1 Breastfeeding

Breastfeeding is still a challenge to date. Mothers have various reasons they are not breastfeeding exclusively namely job/work and insufficient breast milk.

In this paper, findings revealed that caregivers experienced challenges when it comes to breastfeeding. Mothers were unable to breastfeed their children during the day due to work or school. Hence, they only breastfed at night. Children were not breastfed on demand because the mother was tired from schoolwork or work and wakes up early in the morning to go to work and come back late. This is illustrated in the expression below:

> *"The child is only breastfed at night, since the mother is at school during the day and is a grade 12 student who is studying at night until morning"*
>
> (Grandmother, 61 years old)

Another caregiver also revealed that children are not exclusively breastfed on demand.

"The mother breastfeeds the child at night only, since the mother is working during the day. The child is fed formula milk during the day, which she stopped at 7 months, so alternatively, gave her boiled ultraMelk but also along the way also refused it"

(Grandmother, 58 years old)

On the other hand, grandmothers encouraged the mothers to stop breastfeeding the child because the child is not eating food. Their perception is that breast milk is not more important than food. The quotes below show the reason she stopped breastfeeding her child.

"Yes, I was breastfeeding him exclusively but then started giving him soft porridge (prepared by mixing maize meal and water only) *at around 3 or 4 months if not mistaken because my grandmother feels the child is not satisfied of breast milk "mikando" only. Even so, the child chose breast milk over soft porridge so my grandmother decided that we stop giving breast milk at 7 months, she believed that if I stop breastfeeding him, he will be so hungry he will eat soft porridge"*

(Mother, 33 years old)

3.1.2. Sub-Theme 1.2 Non-Nutritious Food

Most participants have poor food choices, but these food choices of primary caregivers (mothers) were different from the food choices of secondary caregivers (grandmothers).

Mothers were found to be feeding their children junk food, e.g., Danone yoghurt, sweets, puree, and juice for kids, because they felt that it is more nutritious than feeding the child soft porridge. The following quotes depict their poor food choices. Participants between 20 and 29 years of age fed their children junk food

"The child is not fed soft porridge only; but I also give her Danone yoghurt, sweets, and juice for children"

(Mother, 21 years old)

"She really enjoys potatoes; I mean potato chips, porridge with artchaar, but Danone yoghurt and juice for children is her most favorable food"

(Mother, 20)

Grandmothers feed the children soft porridge in the morning, and although they tried to provide a variety of things like Cremora powdered coffee creamer, sugar, etc. it was not nutritious enough.

"But I don't give her soft porridge (mix maize meal into a boiling water in a pot and stir until its thick) *but add with some sugar only; I sometimes add "Cremora" powdered coffee creamer to soft porridge since I can't afford to buy formula. But I think I'm feeding him quite right since I'm not giving him soft porridge with sugar only but with Cremora powdered coffee creamer"*

(Grandmother, 49 years old)

"I give the child soft porridge every morning, then pap with vegetables, sometimes with mango artchaar"

(Grandmother, 59 years old)

3.2. *Theme 2: Knowledge Deficit Regarding Signs of Malnutrition*

Participants also reflected having difficulties in identifying a child with malnutrition. This was identified through a child's physical appearance.

Sub-Theme 2.1 Physical Appearance

Recognizing a child with signs and symptoms of malnutrition is difficult. In this paper, few participants referred to the child with malnutrition as healthy but having other conditions like diarrhea. This statement was supported by the following quotation:

"Nurse, I'm surprised that you are saying my grandchild is having Kwashiorkor, really its confusing me, because I can see she is healthy "shining" and fat, so where does this Kwash come from? My grandchild is only passing loose stools and "Ngoma" (fontanel) which is not closing but getting bigger"

(Grandmother, 66 years old)

Inability to identify malnutrition in the child leads to improper treatment. Hence, the caregiver went to a traditional healer to seek help as she believed the child was having diarrhea, which is caused by a "Ngoma" fontanel that is too big and not closing.

"I also took the child to traditional healer, because the traditional healer knows best about "Ngoma" fontanel that is not closing, she is her traditional healer after all. She will know what to do to help this child since" Ngoma" fontanel is not closing; it is getting bigger, so when I arrive, she gave me some "muthi" traditional medicine to apply on the head"

(Grandmother, 69 year old)

She went to the nearest clinic for treatment of diarrhea, because she understands it is a better place to be treated at. That's when she was told about malnutrition which she was not aware of. She used treatment from the clinic and that from a traditional healer. The quotes show how sure she is that both western and traditional medicine work.

"Ok does this means you don't know that your western medications are made from the trees found in the bush, it is just that this white people cooked it, in Tshivenda the traditional healer, burns and pounds them to make "muuluso", a pounded traditional medicine used to protect the child from illness, haven't you heard about "muuluso"

(Grandmother, 69 years old)

4. Discussion

The study findings revealed that caregivers experienced challenges when it comes to breastfeeding. Mothers were unable to breastfeed their children during the day due to work or school, they only breastfed at night. Prashanth et al.; Pravana et al. [18,19] indicated that 42.7% of children were not exclusively fed breast milk for up to six months. Furthermore, it was revealed that children who started with complementary foods before six months, and too late, are likely to develop severe acute malnutrition. However, there are still mothers who know the benefits of breastfeeding and practice it. The study by Ansuya et al. [20] found that out of 58.9% of mothers who practiced exclusive breastfeeding, only 1.89% were at risk of being malnourished.

In this study, mothers between the age of 20 and 29 years fed their children Danone yoghurt, sweets and juice for children. It has resulted in a high rate of children who are underfed among the needy population because food is expensive and they can't afford it [21]. Studies in Karnataka and in Nepal, by Prashanth et al. and Pravana et al. [18,19] show that severe acute malnutrition is common in children born to mothers who are below 20 years. Furthermore, severe acute malnutrition is 3.21 times higher in children born to mothers below 20 years and ≥35 years compared to those mothers aged between 20–34 years. This statement support Table 1, which indicates quite a few children with malnutrition are under the care of caregivers ≥ 35 years of age. This is supported by Drammeh et al. [22] who found that children under the care of teenage mothers or those younger generations are more likely to be undernourished.

While young mothers were giving junk food, grandmothers fed the children soft porridge [23]. Mweemba [23] revealed that not giving enough food, giving food with no nutritious value, and a lack of variety in diet are contributory factors to malnutrition.

Few participants referred to their malnourished children as healthy but having other conditions such as diarrhea. Inability to identify malnutrition in the child leads to improper treatment. In one case, the caregiver went to a traditional healer to seek help as she believed the child was having diarrhea, which is caused by a big fontanel. Mweemba [23] conducted a study in Lusaka and they also confirmed that symptoms of kwashiorkor were confused with other illnesses such as diarrhea. Participant confessed that kwashiorkor is difficult to assess as the child looked fat. This is in line with the findings of this study as caregivers referred a child's unhealthy body to a healthy one. Similarly, Mohseni et al. [12] indicated that knowledge mothers have about their children's health has a significant impact on the children's nutrition level.

5. Conclusions

The study findings describe the perceptions of caregivers regarding malnutrition. Caregivers perceived that the initiation of complementary food early, and not breastfeeding exclusively, were the factors that contributed most to malnutrition. This study will give detail to policy developers and other concerned bodies to combat malnutrition by coming up with appropriate nutritional educational programs including breastfeeding which will reduce the child mortality rate. Furthermore, health care workers could then provide relevant health education which will help in preventing malnutrition and promoting growth.

Author Contributions: T.E.T. assisted with proposal development, was involved in data collection and interpretation, participated in the drafting of the article, and critically read and revised the article. T.C.M., M.M. and N.S.R. supervised the study, proofread the article, and approved the themes. All authors have read and agreed to the published version of the manuscript.

Funding: The University of Venda research publication committee provided financial support. Cost center number: 0253; Account number: 3544.

Institutional Review Board Statement: The study was approved by the University of Venda ethics committee and the Department of Health, Limpopo Province (FHS/21/PDC/25/1301 and 15/05/2018).

Informed Consent Statement: All participants gave consent to participate in the research project voluntarily.

Data Availability Statement: Data sharing will be made available to other researchers on request within the regulations of the University of Venda.

Acknowledgments: The gratitude of this study goes to the University of Venda research ethics committee, for approval and financial support as well as all the participants for taking part in the study.

Conflicts of Interest: The authors declare no conflict of interest.

References

1. Children's Act. *Act No 38 of 2005 as Amended*; Government Printers: Pretoria, South Africa, 2005.
2. South African Government. The constitution of the Republic of South Africa [Homepage on the Internet]. 2021. Available online: https://www.gov.za/documents/constitution-republic-south-africa-1996 (accessed on 17 September 2022).
3. Bongkiynuy, N.; Cumber, S.N.; Jaila, S.; Tsoka-Gwegweni, J.M. Poor complementary feeding practices among young children in Cameroon. *S. Afr. J. Clin. Nutr.* **2017**, *30*, 37–38.
4. Worthman, C.M.; Tomlinson, M.; Rotheram-Borus, M.J. When can parents most influence their child's development? Expert knowledge and perceived local realities. *Soc. Sci. Med.* **2016**, *154*, 62–69. [CrossRef] [PubMed]
5. Vorster, H.H.; Badham, J.B.; Venter, C.S. Food Based Dietary Guidelines for South Africa. *S. Afr. J. Clin. Nutr. Suppl.* **2013**, S1–S164.
6. Vorster, H.; Badham, J.; Venter, C. An introduction to the revised food-based dietary guidelines for South Africa. *S. Afr. J. Clin. Nutr.* **2013**, *26*, S5–S12.
7. Vollmer, S.; Harttgen, K.; Kupka, R.; Subramanian, S.V. Levels and trends of childhood undernutrition by wealth and education according to a composite index of anthropometric failure: Evidence from 146 Demographic and Health Surveys from 39 countries. *BMJ Glob. Health* **2017**, *2*, 4–5. [CrossRef] [PubMed]
8. Department of Social Development, South African Social Security Agency & Unicef. *Child Support Grant Evaluation 2010: Qualitative Research Report*; UNICEF: Pretoria, South Africa, 2011.
9. World Health Organization. Levels and Trends in Child Malnutrition: UNICEF/WHO/The World Bank Group Joint Child Malnutrition Estimates: Key Findings of the 2020 Edition. In *Joint Child Malnutrition Estimates*. World Health Organization: Geneva, Switzerland; Washington, DC, USA, 2020.
10. May, J.; Witten, C.; Lake, L. The slow violence of malnutrition. In *South African Child Gauge 2020*; May, J., Witten, C., Lake, L., Eds.; Children's Institute, University of Cape Town: Cape Town, South Africa, 2020; pp. 24–45.
11. Murad, A.; Renfrew, M.J.; Symon, A.; Whitford, H. Understanding factors affecting breastfeeding practices in one city in the Kingdom of Saudi Arabia: An interpretative phenomenological study. *Int. Breastfeed. J.* **2021**, *16*, 9. [CrossRef] [PubMed]

12. Mohseni, M.; Aryankhesal, A.; Kalantari, N. Factors associated with malnutrition among under five-year-old children in Iran: A systematic review. *Ann. Trop. Med. Public Health* **2017**, *10*, 1147–1158.
13. Omotoye, F.E.; Adesanmi, R.A.S. Infant and young child-feeding practices in two local government areas in Southwest, Nigeria. *J. Food Sci. Nutr. Res.* **2019**, *2*, 136–145. [CrossRef]
14. Mushaphi, L.F.; Mahopo, T.C.; Nesamvuni, C.N.; Baloyi, B.; Mashau, E.; Richardson, J.; Dillingham, R.; Guerrant, R.; Ambikapathi, R.; Bessong, P. Recommendations for infant feeding policy and programs in Dzimauli region, South Africa: Results from the MAL-ED birth cohort. *Food Nutr. Bull.* **2017**, *38*, 428–440. [CrossRef] [PubMed]
15. Motadi, S.A.; Malise, T.; Mushaphi, L.F. Breastfeeding knowledge and practices among mothers of children younger than 2 years from a rural area in the Limpopo Province, South Africa. *S. Afr. J. Child. Health* **2019**, *13*, 115–119. [CrossRef]
16. Lincoln, Y.S.; Guba, E.G. *Naturalistic Inquiry*; Sage Publications: Newbury Park, CA, USA, 1985.
17. Creswell, J.W. *Research Design, Qualitative, Quantitative and Mixed Methods Approaches*, 4th ed.; Department of Family Medicine, University of Michigan, Sage Publishing: Thousand Oaks, CA, USA, 2014.
18. Prashanth, M.R.; Savitha, M.R.; Prashanta, B. Risk factors for severe acute malnutrition in under five children attending nutritional rehabilitation centre of a tertiary teaching hospital in Karnataka: A case-control study. *Int. J. Contempo. Pediatr.* **2017**, *4*, 1721–1726. [CrossRef]
19. Pravana, N.K.; Piryani, S.; Chaurasiya, S.P. Determinants of severe acute malnutrition among children under 5 years of age in Nepal: A community base case- control study. *BMJ Open* **2017**, *7*, e017084. [CrossRef] [PubMed]
20. Nayak, B.S.; Unnikrishnan, B.; George, A.; Shashidhara, Y.N.; Mundkur, S.C.; Guddattu, V. Risk factors for malnutrition among preschool children in rural Karnataka: A case-control study. *BMC Public Health* **2018**, *18*, 283.
21. Koetaan, D.; Smith, A.; Liebenberg, A.; Brits, M.; Halkas, C.; Van Lill, M.; Joubert, G. The prevalence of underweight in children aged 5 years and younger attending primary health care clinics in the Mangaung area, Free State. *Afr. J. Prim. Health Care Fam. Med.* **2018**, *10*, 5. [CrossRef] [PubMed]
22. Drammeh, W.; Hamid, N.A.; Rohana, A.J. Determinants of household food insecurity and its association with child malnutrition in Sub-Saharan Africa: A review of the literature. *Curr. Res. Nutr. Food Sci.* **2019**, *7*, 610–623. [CrossRef]
23. Mweemba, N. Mother's perception of protein energy malnutrition in children under five years: An exploratory and descriptive qualitative study conducted at Chipata Health center, Chipata compound in Lusaka. *Glob. Res. J. Public Health Epidemiol.* **2017**, *4*, 87–97.

children

Article

Overweight Prevalence among Rural Adolescents by Household Head Obesity and Socio-Economic Status in Limpopo, South Africa

Peter M. Mphekgwana [1,*], Masenyani O. Mbombi [2], Livhuwani Muthelo [2], Joseph Tlouyamma [3], Rathani Nemuramba [3], Cairo Ntimana [3], Katlego Mothapo [3], Inos Dhau [4,5] and Eric Maimela [3,5]

1 Research Administration and Development, University of Limpopo, Sovenga St, Polokwane 0727, South Africa
2 Department of Nursing Science, University of Limpopo, Sovenga St, Polokwane 0727, South Africa
3 DIMAMO Population Health Research Centre, University of Limpopo, Sovenga St, Polokwane 0727, South Africa
4 Department of Geography and Environmental Studies, University of Limpopo, Sovenga St, Polokwane 0727, South Africa
5 Department of Public Health, University of Limpopo, Sovenga St, Polokwane 0727, South Africa
* Correspondence: peter.mphekgwana@ul.ac.za

Abstract: Background: Childhood obesity has become the most important and growing public health problem in the world. They add to public health challenges by increasing the burden of chronic non-communicable diseases. However, in spite of its importance, there is limited literature that evaluates the prevalence of obesity among rural adolescents in sub-Saharan Africa. We report the first study to present an insight into rural black overweight South African children and the physical characteristics and socio-economic status of the household head. A quantitative cross-sectional population study was conducted involving 51 selected primary villages within the DIMAMO surveillance area in the Capricorn District of Limpopo Province, South Africa. The study involved 294 adolescents, 154 girls and 140 boys, who were under the age of 18. Of these participants, 127 (43%) were within the normal weight range, and 167 (57%) were overweight. Gender made a significant difference, with more girls being overweight than boys. Adolescents who did not receive child grants and whose heads of household were 45–54 years old, poor, and overweight had a higher prevalence of obesity (p-value < 0.05). This study suggests that public health interventionists need to target both the heads of household and their children in hopes of reducing the prevalence of overweight and obese South African children. We further propose a better understanding of the causes of childhood overweight and obesity to guide policy development and implementation in rural settings.

Keywords: overweight; household head; blood pressure; adolescents; rural

1. Introduction

Being overweight and obese has become the most important and growing public health concern for children around the world. They add to the public health challenge by increasing the burden of chronic non-communicable diseases (NCDs), which affect both developing and developed countries, including South Africa [1,2]. The prevalence amongst children aged 5 and 19 years has increased from 30.3 million in 2000 to 38.3 million in 2019 [3]. It is a multifactorial condition marked by excess adiposity caused by a complex interaction of biological, developmental, genetic, and environmental factors [4–6]. Obesity and overweight in children are on the rise in low- and middle-income countries, particularly in urban areas, with about a quarter of African children under the age of five being overweight. The primary cause of this in children is diet-related factors, including unhealthy food choices, such as high-energy foods, and also a lack of physical activity and/or a sedentary lifestyle [4,7–9].

The significant effects of obesity on children are grouped as physical health, social, psychological health, educational, and emotional well-being, such as developing early puberty in children, menstrual irregularities in adolescent girls, sleep disorders such as poor self-esteem, body image, and peer relationships, eating disorders, poor academic performance, and a lower quality of life [10]. Furthermore, Nicolucci and Maffeis (2022) noted that obesity in children represents a strong predictor of the condition and higher mortality in adulthood [11]. Moreover, children who grow up in poverty are 1.5 times more likely to be overweight and 1.6 times more likely to be obese [12], and the adverse effects of poverty and being overweight or obese have a far-reaching impact on educational outcomes.

Given the above-mentioned significant effect of obesity on children, there is a need for implementing effective treatments. So far, the effectiveness of obesity interventions has been based on health screening, individual behavioral changes, increasing daily physical exercise, or improving the quality of diet by restricting excess calorie intake [4,11,13]. For instance, health screening, a medical procedure performed to assess the likelihood of having a particular problem such as obesity in children [14], is used to determine if a child is obese or not by studying their BMI. On the other hand, prevalence studies could be used to inform researchers, guideline developers, and policymakers about the burden of obesity and the effectiveness of treatment for a targeted population [15].

However, there is limited literature in sub-Saharan Africa (SSA) to assess the prevalence of obesity among rural children by education level, employment, and the obesity status of the head of household in the rural setting. This burden of obesity in children is not foreign to South Africa, where the prevalence of overweight in children (aged 14–17) of African ancestry has been reported to be between 30% and 40% [16]. In the study reported here, we surveyed 294 adolescents with their heads of households to discover characteristics of Limpopo household heads related to obesity in rural-dwelling black South African children. We sought to deepen our understanding of the causes of being overweight and obese in children in order to guide policy development and the implementation of interventions to reduce the prevalence of overweight in rural children.

2. Methodology

2.1. Study Design

A quantitative cross-sectional population study was conducted within a bigger ongoing project of the DIMAMO Health and Demographic Surveillance Site (HDSS) involving 51 selected primary villages within the Dikgale, Mamabolo, and Mothiba (DIMAMO) surveillance area in the rural Capricorn District of Limpopo Province, South Africa. There were about 100,000 people under surveillance, mostly Northern Sotho speakers. However, there were few adolescents in these households. The study involved 294 adolescents, 154 girls and 140 boys, who were under the age of 18. Each household within the study site received a structured questionnaire that was developed, incorporated into the survey solutions, and distributed by trained field workers. All parents or guardians of the participants consented and the participants gave their informed consent in writing, while the Turfloop Research Ethics Committee (TREC) committees approved the study's protocol. The tribal authority within the study setting authorized the study's execution and provided written, informed consent.

2.2. Measurement

According to the International Society for the Advancement of Kinanthropometry protocol, each child underwent a series of anthropometric measurements [17]. With the subject in an anatomical position and the head in a Frankfort plane, height was measured with a Martin anthropometer to the nearest 0.1 cm. On electronic scales, bodyweight was calculated to the nearest 0.1 kg while wearing light clothing and no shoes. Body mass index (BMI) was determined by dividing weight in kilograms per square meter (kg/m^2) by height. All anthropometric measurement training was carried out following the ISAK's established protocols [17]. Demographic data for the participant, including age, gender, and educational attainment, were taken from the Dikgale HDSS database. The questionnaire was initially created in English, translated into the local language of Northern Sotho, and validated using construct, content, and face validity, as well as a pilot study.

2.3. Statistical Analyses

Statistical Package for Social Sciences (SPSS) version 26.0 software (I.B.M., Armonk, New York, NY, USA) was used to compute descriptive and inferential statistics. Data were reported as frequency and percentages. A comparison of categorical variables was performed using chi-square, and the statistical significance was set at $p < 0.05$ with a 95% confidence interval. Logistic regression analyses were used to determine the association between overweight adolescents with selected socioeconomic covariates and overweight status among the head of household.

3. Results

The characteristics of the participants are described in Table 1. Of the 294 participants, 127 (43.2%) were within the normal weight limit, and 167 (56.8%) were overweight. Most heads of household were under 30 years old (35%), followed by those aged ≥55 years. There were almost twice as many female (68%) household heads as males. More than half of household heads said they were just scarping by financially (65%) and more than 80% obtained less than a high school and were unemployed. Most adolescents were receiving child grants and living with their relatives. Around fifty-seven percent (57%) of household heads were overweight (Table 1).

Table 1. Characteristics of the participants by BMI group (normal and overweight).

	Total (%)	Normal Weight (%) Adolescents 127 (43)	Overweight (%) Adolescents 167 (57)	Chi-Square *p*-Value
Child's (gender)				<0.001 *
Girl	154 (52)	106 (69)	48 (31)	
Boy	140 (48)	131 (94)	9 (6)	
Social grant				0.001 *
No	48 (16)	30 (62)	18 (38)	
Yes	246 (84)	207 (84)	39 (16)	
Head of household (gender)				0.943
Female	200 (68)	161 (81)	39 (19)	
Male	94 (32)	76 (81)	18 (19)	
Age group				0.016 *
19–34	103 (35)	81 (79)	22 (21)	
35–44	59 (20)	53 (90)	6 (10)	
45–54	52 (18)	35 (67)	17 (33)	
55+	80 (27)	68 (85)	12 (15)	
Financial status				0.005 *
Comfortable	29 (10)	22 (76)	7 (24)	
Extremely poor	12 (4)	12 (100)	0 (0)	
Just getting by	183 (65)	156 (85)	27 (15)	
Poor	56 (20)	30 (64)	26 (36)	
Education level				0.485
Above high school	29 (10)	24 (83)	5 (17)	
High school	5 (2)	3 (60)	2 (40)	
Less than high school	260 (88)	210 (81)	50 (19)	
Employment status				0.802
No	239 (81)	192 (80)	47 (20)	
Yes	55 (19)	45 (82)	10 (18)	

Table 1. *Cont.*

	Total (%)	Normal Weight (%) Adolescents 127 (43)	Overweight (%) Adolescents 167 (57)	Chi-Square *p*-Value
Religious faith				0.716
No	53 (24)	46 (87)	7 (13)	
Yes	164 (76)	139 (85)	25 (15)	
	Head of household overweight			<0.001 *
No	127 (43)	117 (92)	10 (8)	
Yes	167 (57)	120 (72)	47 (28%)	
Relationship				0.601
Parents	78 (27)	65 (83)	13 (17)	
Relatives	207 (70)	164 (79)	43 (21)	
Caretaker	9 (3)	8 (89)	1 (11)	

* 0.05 significant level.

There are no significant differences between the overweight status of the child and the prevalence of household head gender, education level, employment status, religious faith, and relationship with the child (*p*-value > 0.05). There were significant gender differences in terms of the prevalence of the child's weight being within the normal range or overweight, with more girls being overweight than boys. Table 1 shows that adolescents who were not receiving a child grant, had an overweight head of household aged 45–54 years, and were poor had a higher prevalence of being overweight (*p*-value < 0.05).

The following factors were identified in the logistic regression model as being associated with overweight adolescents: (a) child's sex, with boys having a lower risk than girls do (OR = 0.163; 95% CI: 0.067–0.396); (b) age of household head: adolescents with a household head aged between 45 and 54 years are at higher risk of being overweight than adolescents with a household head aged over 54 years (OR = 3.720; 95% CI: 1.203–11.501). In comparison to those who did not receive a child grant, those who received child grants had a lower risk of being overweight. The higher odds of adolescents who were overweight were found in households where the household head was overweight (Table 2).

Table 2. Logistic regression model for the overweight adolescents with selected socioeconomic covariates among the heads of households.

	OR	95%CI	*p*-Value
Head of household: Male	1.411	(0.562; 3.539)	0.463
Child: Boy	0.163	(0.067; 0.396)	<0.001 *
Age group			
19–34	2.364	(0.885; 6.313)	0.086 **
35–44	0.897	(0.243; 3.307)	0.871
45–54	3.720	(1.203; 11.501)	0.023 *
55+	Reference		
Financial status			
Just getting by	0.333	(0.098; 1.127)	0.077 **
Poor	0.830	(0.218; 3.164)	0.785
Comfortable	Reference		
Education status			
High school and above	1.068	(0.353; 3.226)	0.908
Less than high school	Reference		
Unemployed	1.473	(0.490; 4.424)	0.491
Receiving child grants	0.281	(0.112; 0.703)	0.007 *
Head of household overweight	5.757	(2.316; 14.312)	<0.001 *

* Significant at 5% level; ** Significant at 10% level.

The mean values of systolic and diastolic blood pressure with respect to age are shown in Figures 1 and 2, respectively. As can be seen, mean systolic blood pressure was higher in adolescents who were overweight. Systolic blood pressure also decreased with age in both normal weight and overweight groups. However, there were no significant difference in the mean blood pressure (systolic and diastolic) between the normal weight and overweight groups.

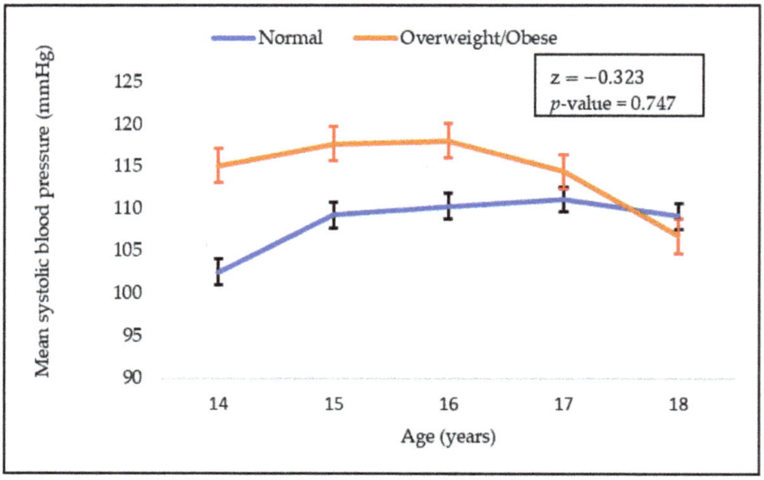

Figure 1. Adolescent's mean systolic blood pressure values according to their BMI group (normal and overweight).

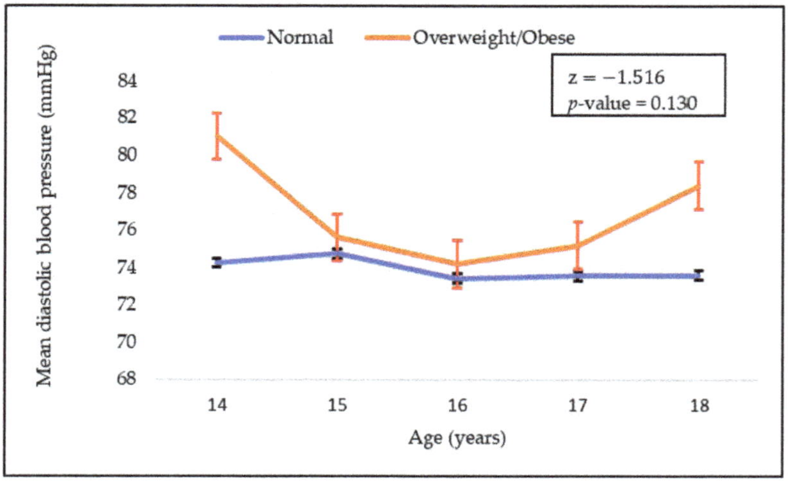

Figure 2. Adolescent's mean diastolic blood pressure values according to their BMI group (normal and overweight).

4. Discussion

Sustainable development goal three in South Africa emphasizes establishing good health and well-being, including the health of children and adolescents. Hence, this study focused on the prevalence of obesity and its components among adolescents in the Dikgale Health and Demographic Surveillance System of Limpopo, South Africa (HDSS). The study further elucidated the odds ratio of overweight adolescents with selected variables of the household head. The prevalence of overweight children was 57%, with a high prevalence observed among those who are receiving child grants. Child obesity alone was reported to be less than 22.8% among South African children in 2010 [18]. The results of the current study show a rise in overweight rural children in the country. This is consistent with a recent study by Engwa et al. (2022) that revealed an increase in the prevalence of overweight in children [16]. However, the prevalence was high compared to studies in other sub-Saharan African countries such as Sierra Leone (16.9%), Comoros (15.9%), Malawi (14.5%), Ethiopia (3.0%), Togo (2.6), and Senegal (2.0%) [19]. The rise in prevalence is fuelled by etiological factors such as genetic, environmental, socio-economic risk, parental obesity, and parental income [20,21].

Child obesity is reported to differ between males and females [11,16,22]. The prevalence of being overweight was previously reported to be higher in girls than boys during adolescence, but there was no gender difference in early childhood [22]. This result is consistent with our findings, which showed that there were more girls than boys who were overweight and indicated that boys have a lower risk of being overweight than girls. This situation can be seen in the adult population as well, with women having the greatest likelihood of being overweight [23,24]. A study by Zong and his colleagues found that children raised in an extended family, mainly raised by their grandparents, or in high-income households may have an increased risk of becoming obese/overweight [25]. Contrary to other research [25], this study shows that adolescents staying with household heads between the ages of 45 and 54 are more likely to be overweight than adolescents living with household heads over the age of 54. The results' discrepancy may be due to the socioeconomic disparities between the two nations.

A study by Shiba and Kondo looked into whether differences in the risk of becoming overweight between Japanese children from single-parent and two-parent families rose following the 2008 financial crisis [26]. In their study, they discovered that during the 2008 financial crisis, the risk of being overweight increased significantly for children from single-parent households. As a result, it was concluded that financial support and more childrearing support are needed to provide opportunities for single parents seeking jobs, social relationships, and support in raising their children. The current study showed a significant relationship between child grants and the weight status of children, with more adolescents who did not receive child grants being overweight. The Child Support Grant (CSG) is an important instrument of social protection in South Africa, reaching over 10 million South African children each month [27]. The South African CSG was first introduced in 1998. Early enrolment in the CSG reduced the likelihood of illness, with the effect being particularly strong for boys, particularly. It was further reported that receipt of the grant by adolescents generates a range of positive impacts, not the least of which is a reduction in risky behaviors [27].

Our results revealed that there was no significant difference in the systolic and diastolic blood pressure of normal weight and overweight children. However, systolic blood pressure was higher in adolescents who were overweight, and it decreased with age. Higher odds of overweight adolescents were found in households where the head of the household was overweight. A parent's weight status has a significant influence on the risk of childhood obesity, and it was suggested that parents might play a crucial role in mitigating child obesity [2]. Underaged children are generally dependent on their parents' choices of food. This is the case whether the food is healthy or not. Thus, the consumption of energy-dense food by parents may suggest the same for their children. Energy-dense food is one of the main contributors to obesity [20]. Parents lifestyle choices, such as a sedentary lifestyle and lack of physical activity, can influence those of their children. The type of food consumed by families is influenced by their lifestyle, affordability, and socio-economic status.

Childhood obesity associated with household socio-economic status such as low-income households was reported to be the leading factor of obesity among children [28–32]. The relationship between a parent's or household's socio-economic status and child obesity was stronger in Asia than in Europe and the Middle East, and it was stronger in high-income countries than in middle- or low-income countries [2]. However, our study, which was conducted in middle- or low-income countries, found no significant association between childhood obesity and the head of household socio-economic factors such as employment. The disparity could be attributed to the study's location as well as the small sample size. In addition, the educational level of the head of the household was found not to be statistically associated with childhood overweight. This contradicted previous findings, which stated that parents' educational level might contribute to their children's nutritional knowledge, which may lead to the development of obesity in their children [28,33,34]. This might be because the current study had a majority (88%) of the heads of the household who completed less than high school, which could have led to a discrepancy in the study findings.

Study Limitations

The study included the following possible limitations: First, the study was a cross-sectional design, preventing the assertion of a causal association between anthropometric indices of adiposity and blood pressure. The second factor is the small patient population.

5. Conclusions

Obesity in children has been attracting increasing attention in rural and urban areas because of concerns about its long-term effects on adulthood. The present study presents a rare insight into rural-dwelling black South African children and adolescents who are overweight and the influence of household head socioeconomic status on their overweight. The results of the present study suggest that variables such as household head gender, education level, employment status, religious faith, and relationship with the child are not associated with child overweight in rural areas. Children may inherit genetic predispositions from their parents. In an area dominated by cardiovascular risk factors (i.e., overweight or obesity and hypertension), screening and monitoring children's anthropometric indices and blood pressure regularly is needed for the effective prevention and management of chronic diseases. We further propose that public health interventionists be targeted both at heads of household and children in hopes of reducing the prevalence of obesity in South Africa's children.

Author Contributions: P.M.M., M.O.M., L.M., J.T., R.N., C.N., K.M., I.D. and E.M. contributed equally to this manuscript. All authors have read and agreed to the published version of the manuscript.

Funding: This research was funded by the South African Population Research Infrastructure Network (SAPRIN).

Institutional Review Board Statement: The study was approved by the University of Limpopo Research Ethics Committee (TREC/58/2020: PG) on 24 April 2020. The Tribal Authority within the study setting authorized the study's execution and provided written informed consent forms were completed by each participant.

Informed Consent Statement: The tribal authority within the study setting authorized the study's execution and provided written, informed consent. Informed consent was obtained from all subjects involved in the study and the tribal authority within the study setting authorized the study's execution and provided written, informed consent.

Data Availability Statement: The data presented in this study are available on request from the corresponding author. The data are not publicly available due to privacy or ethical restrictions.

Acknowledgments: This study was conducted as part of the South African Population Research Infrastructure Network (SAPRIN), an initiative hosted by the South African Medical Research Council and receives long-term funding from the National Department of Science and Innovation. We acknowledge the long-term financing of the SAPRIN, which made it possible for DIMAMO to collect data in the identified rural areas continuously.

Conflicts of Interest: The authors declare no conflict of interest.

References

1. Qasim, A.; Turcotte, M.; De Souza, R.J.; Samaan, M.C.; Champredon, D.; Dushoff, J.; Speakman, J.R.; Meyre, D. On the origin of obesity: Identifying the biological, environmental and cultural drivers of genetic risk among human populations. *Obes. Rev.* **2018**, *19*, 121–149. [CrossRef] [PubMed]
2. Lin, X.; Li, H. Obesity: Epidemiology, pathophysiology, and therapeutics. *Front. Endocrinol.* **2021**, 1070. [CrossRef] [PubMed]
3. Zatońska, K.; Psikus, P.; Basiak-Rasała, A.; Stępnicka, Z.; Gaweł-Dąbrowska, D.; Wołyniec, M.; Gibka, J.; Szuba, A.; Połtyn-Zaradna, K. Obesity and Chosen Non-Communicable Diseases in PURE Poland Cohort Study. *Int. J. Environ. Res. Public Health* **2021**, *18*, 2701. [CrossRef] [PubMed]
4. Templin, T.; Hashiguchi, T.C.O.; Thomson, B.; Dieleman, J.; Bendavid, E. The overweight and obesity transition from the wealthy to the poor in low- and middle-income countries: A survey of household data from 103 countries. *PLoS Med.* **2019**, *16*, e1002968. [CrossRef]
5. World Health Organization. Obesity and Overweight. Available online: https://www.who.int/news-room/fact-sheets/detail/obesity-and-overweight (accessed on 20 June 2022).
6. Beas, C. Assessing Nutrition Barriers and Childhood Obesity Factors among Children with Autism Spectrum Disorder. Master's Thesis, California State University, San Marcos, CA, USA, 2020.
7. James, P.T. Obesity: The worldwide epidemic. *Clin. Dermatol.* **2004**, *22*, 276–280. [CrossRef]
8. Scaglioni, S.; De Cosmi, V.; Ciappolino, V.; Parazzini, F.; Brambilla, P.; Agostoni, C. Factors influencing children's eating behaviours. *Nutrients* **2018**, *10*, 706. [CrossRef]
9. Wang, Y.; Beydoun, M.A. The Obesity Epidemic in the United States Gender, Age, Socioeconomic, Racial/Ethnic, and Geographic Characteristics: A Systematic Review and Meta-Regression Analysis. *Epidemiol. Rev.* **2007**, *29*, 6–28. [CrossRef]
10. Chong, M.F.-F.; Lim, H.X.; Wong, B.W.X.; Chi, Z.H.; Inthujaa, J.K.; Müller-Riemenschneider, F.; Cheon, B.K.; Gorny, A.W.; Chia, K.S. Transiting out of Full-Time National Service: A Qualitative Study of Barriers and Motivators of Weight Change in Young Adult Men in Singapore. *Am. J. Men's Health* **2022**, *16*. [CrossRef]
11. Manca, R.; Bombillar, F.; Glomski, C.; Pica, A. Obesity and immune system impairment: A global problem during the COVID-19 pandemic. *Int. J. Risk Saf. Med.* **2022**, *33*, 193–208. [CrossRef]
12. Mchiza, Z.J.-R.; Parker, W.-A.; Hossin, M.Z.; Heshmati, A.; Labadarios, D.; Falkstedt, D.; Koupil, I. Social and Psychological Predictors of Body Mass Index among South Africans 15 Years and Older: SANHANES-1. *Int. J. Environ. Res. Public Health* **2019**, *16*, 3919. [CrossRef]
13. Modjadji, P. Socio-demographic Determinants of Overweight and Obesity among Mothers of Primary School Children Living in a Rural Health and Demographic Surveillance System Site, South Africa. *Open Public Health J.* **2020**, *13*, 518–528. [CrossRef]
14. Mashinya, F.; Alberts, M.; Cook, I.; Ntuli, S. Determinants of body mass index by gender in the Dikgale Health and Demographic Surveillance System site, South Africa. *Glob. Health Action* **2018**, *11*, 1537613. [CrossRef]
15. Nicolucci, A.; Maffeis, C. The adolescent with obesity: What perspectives for treatment? *Ital. J. Pediatr.* **2022**, *48*, 9. [CrossRef]
16. Engwa, G.A.; Anye, C.; Nkeh-Chungag, B.N. Association between obesity and lung function in South African adolescents of African Ancestry. *BMC Pediatr.* **2022**, *22*, 109. [CrossRef]
17. Tanner, J.M.; Whitehouse, R.H.; Takaishi, M. Standards from birth to maturity for height, weight, height velocity, and weight velocity: British children, 1965. *I. Arch. Dis. Child.* **1966**, *41*, 454–471. [CrossRef]
18. Skinner, A.C.; Perrin, E.M.; Skelton, J.A. Prevalence of obesity and severe obesity in US children, 1999–2016. *Pediatrics* **2016**, *141*, e20173459. [CrossRef]
19. Mkuu, R.; Barry, A.; Yonga, G.; Nafukho, F.; Wernz, C.; Gilreath, T.; Chowdhury, M.A.; Harvey, I.S. Prevalence and factors associated with overweight and obesity in Kenya. *Prev. Med. Rep.* **2021**, *22*, 101340. [CrossRef]
20. Elkum, N.; Alarouj, M.; Bennakhi, A.; Shaltout, A. The Complex Etiology of Childhood Obesity in Arabs Is Highlighted by a Combination of Biological and Socio-Economic Factors. *Front. Public Health* **2019**, *7*, 72. [CrossRef]
21. Mphekgwana, P.M.; Monyeki, K.D.; Makgopa, H.M.; Makgae, P.J. Multiple Points Change in the Association of Blood Pressure Subtypes with Anthropometric Indices of Adiposity among Children in a Rural Population. *Children* **2020**, *7*, 28. [CrossRef]
22. Otitoola, O.; Oldewage-Theron, W.; Egal, A. Prevalence of overweight and obesity among selected schoolchildren and adolescents in Cofimvaba, South Africa. *S. Afr. J. Clin. Nutr.* **2021**, *34*, 97–102. [CrossRef]
23. Muhammad, T.; Boro, B.; Kumar, M.; Srivastava, S. Gender differences in the association of obesity-related measures with multi-morbidity among older adults in India: Evidence from LASI, Wave-1. *BMC Geriatr.* **2022**, *22*, 171. [CrossRef] [PubMed]
24. Li, J.-B.; Qiu, Z.-Y.; Liu, Z.; Zhou, Q.; Feng, L.-F.; Li, J.-D.; Zhang, X. Gender Differences in Factors Associated with Clinically Meaningful Weight Loss among Adults Who Were Overweight or Obese: A Population-Based Cohort Study. *Obes. Facts* **2021**, *14*, 108–120. [CrossRef] [PubMed]
25. Zong, X.-N.; Li, H.; Zhang, Y.-Q. Family-related risk factors of obesity among preschool children: Results from a series of national epidemiological surveys in China. *BMC Public Health* **2015**, *15*, 927. [CrossRef] [PubMed]
26. Shiba, K.; Kondo, N. The Global Financial Crisis and Overweight among Children of Single Parents: A Nationwide 10-Year Birth Cohort Study in Japan. *Int. J. Environ. Res. Public Health* **2019**, *16*, 1001. [CrossRef] [PubMed]
27. UNICEF. *The South African Child Support Grant Impact Assessment: Evidence from a Survey of Children, Adolescents and Their Households*; UNICEF: New York, NY, USA, 2012.

28. Mahmoodi, Z.; Gill, P.; Qorbani, M.; Khonsari, N.M.; Sheidaei, A.; Heshmat, R.; Heidari-Beni, M.; Kelishadi, R. Socioeconomic inequality in different phenotypes of childhood obesity and its determinants in Iran: A Blinder-Oaxaca decomposition method. *BMC Public Health* **2022**, *22*, 1580. [CrossRef]
29. De Onis, M.; Lobstein, T. Defining obesity risk status in the general childhood population: Which cut-offs should we use? *Int. J. Pediatri. Obes.* **2010**, *5*, 458–460. [CrossRef]
30. Gebremedhin, S. Prevalence and differentials of overweight and obesity in preschool children in Sub-Saharan Africa. *BMJ Open* **2015**, *5*, e009005. [CrossRef]
31. Butte, N.F.; Cai, G.; Cole, S.A.; Comuzzie, A.G. Viva la Familia Study: Genetic and environmental contributions to childhood obesity and its comorbidities in the Hispanic population. *Am. J. Clin. Nutr.* **2006**, *84*, 646–654. [CrossRef]
32. Pulgarón, E.R. Childhood Obesity: A Review of Increased Risk for Physical and Psychological Comorbidities. *Clin. Ther.* **2013**, *35*, A18–A32. [CrossRef]
33. Lopez-Cepero, A.; Frisard, C.F.; Lemon, S.C.; Rosal, M.C. Association between emotional eating, energy-dense foods and overeating in Latinos. *Eat. Behav.* **2019**, *33*, 40–43. [CrossRef]
34. Jastreboff, A.M.; Chaplin, T.M.; Finnie, S.; Savoye, M.; Stults-Kolehmainen, M.; Silverman, W.K.; Sinha, R. Preventing Childhood Obesity through a Mindfulness-Based Parent Stress Intervention: A Randomized Pilot Study. *J. Pediatr.* **2018**, *202*, 136–142. [CrossRef]

children

MDPI

Article

Alcohol Use among High School Learners in the Peri-Urban Areas, South Africa: A Descriptive Study on Accessibility, Motivations and Effects

Mmampedi Mathibe [1], Lindiwe Cele [1] and Perpetua Modjadji [1,2,*]

[1] Department of Public Health, School of Health Care Sciences, Sefako Makgatho Health Sciences University, 1 Molotlegi Street, Ga-Rankuwa 0208, South Africa

[2] Non Communicable Diseases Research Unit, South African Medical Research Council, Cape Town 7505, South Africa

* Correspondence: Perpetua.modjadji@mrc.ac.za; Tel.: +27-21-938-0210

Abstract: Learners are vulnerable to alcohol use and its negative effects, largely due to accessibility of alcohol products, especially in the localities with poor socioeconomic status and infrastructure. This study aimed to determine the accessibility, motivations and effects of alcohol use among high school learners (n = 403) in Tshwane North and West, South Africa, using a descriptive, cross-sectional design. Data were collected using a validated questionnaire and analysed using STATA 17. Learners (16 ± 2 years) had poor demographic status and lifestyle behaviors. Availability of alcohol outlets (54%) near schools was reported, and learners indicated easy access (65%) to taverns and bottle stores (30%), and purchasing alcohol without a proof of identity document (70%). Motivations for alcohol use were self-pleasure (36%), coping with stress (24%) and increasing self-esteem (19%). Almost half of the learners (49%) introduced themselves to alcohol use, while others were influenced by friends (36%) and family (14%). Reported alcohol related effects were a negative impact on health (56%), brain function (25%) and school work (12%), in addition to social harms, including problems with friends (25%) and parents (17%), physical fights (19%) and engaging in risky sexual behaviour (11%). Effective strategies are necessary to address underage alcohol use and should include regulating the proximity of alcohol outlets to schools, life skills training to address learners' drinking motives and constantly alerting parents about the relevance of modeling behaviour.

Keywords: alcohol use; accessibility; motivations; effects; high school learners; peri-urban; Tshwane; South Africa

1. Introduction

Alcohol use and abuse is increasingly becoming a worldwide lifestyle that is prevalent in rich and poor countries alike and one of the top three major health problems [1,2]. In sub-Saharan Africa (SSA), the prevalence of substance use among adolescents is approximated at 42%, with alcohol (40.8%) and tobacco (45.6%) being the highest prevailing substances used across Africa continent, compared to other substances [3,4]. In particular, South Africa has growing evidence of risky alcohol use among various population groups [5,6], in addition to tobacco use [7–10]. The country continues to experience considerable change with the onset age of alcohol intake among school age adolescents across the society [11–14]. Studies in South Africa have estimated youth drinking at 32%, especially between the ages of 11–20 years, while 12% of teenagers are reported to initiate alcohol use before 13 years of age [14–16]. Additionally, recent data showed the prevalence of alcohol use among high school learners ranging from 22% to 69% [17–21].

Accessibility and availability of alcohol from various sources increase alcohol use among adolescents, including direct purchasing from the commercial establishments, despite laws prohibiting such sales to young people under the age of 18 years [22], which is also the case in South Africa [23,24]. Literature suggests that alcohol availability is categorised in terms of physical, economic, social and subjective or psychological availability [25]. Physical availability refers to the arrangements made by governments that determine the convenience or difficulty to obtain and consume alcoholic beverages. Economic availability refers to the price and disposable income affecting the affordability of alcohol. Social availability refers to the degree of normative support for drinking provided by a person's key social groups, such as family, friends, sports clubs and the neighbourhood public house. Subjective or psychological availability refers to the perception of accessible people on alcohol and their response to alcohol marketing [26]. Researchers in other countries, including South Africa, have reported an established positive link between the number of liquor outlets found in a locality, as well as proximity, density of alcohol outlets and indulgence in alcohol use in the communities [24,27–31].

A combination of different variables has been implicated to influence the use of alcohol among adolescents [32]. Adolescence is a developmental period of heightened responsiveness to social rewards, and this implies an increasing desire to fit in among peers [33]. Drinking motives and perceived peer and parental norms are strong predictors of alcohol use [34]. Identifying the adolescents' motivations for their alcohol use has revealed that adolescents consume alcohol when they are having fun with their friends. Moreover, adolescents use alcohol to overcome the stressful life events and conflict situations in relationships with their family and friends, while for others it is self-reward after an achievement [35]. Further literature documents that alcohol use during adolescence is influenced by learning to use alcohol based on social observation of other significant persons' behaviours (family members, friends, peers, social models), developing positive attitudes towards alcohol and its effects and a result of social imitation and social modelling processes by peer influences (i.e., social learning), as reported by Benţea [36]. Therefore, personal motivations and perceived social norms influence adolescent alcohol use [34]; it can be deduced that most adolescents consume alcohol as a result of the pleasure, peer group, and family influence [34,36]. In addition to South Africa [17,37], studies in other countries, such as Israel [38], Namibia [28,39], and Spain [36] have confirmed drinking motives and influences of significant persons among adolescents.

Alcohol is an addictive, psychoactive substance that can cause significant harm to the individual if consumed excessively [40]. Alcohol use among adolescent is associated with many serious social and developmental issues and behaviour problems, such as school absenteeism, acts of aggression and violence, suicidal behaviour, delinquency, antisocial behaviours, neuropsychiatric disorders, etc. [36]. Alcohol use disorders in adolescence and later in adulthood are attributed to early initiation of alcohol use among school age adolescents [41,42]. As previously stated, substance use among adolescents can lead to an increased risk of transmission of sexually transmitted infections, juvenile delinquency and other problems associated with physical and mental health [43]. Specifically, alcohol misuse is associated with an increased number of sexual partners, unprotected sex and sexual encounters that are later regretted [44,45]. Morojele and colleagues [46] reported that adolescents who drink are more likely to have early sexual debut and engage in multiple sexual partnerships.

Accessibility of alcohol products in the localities with poor socioeconomic status and infrastructure have been associated with increased alcohol use, coupled with drinking motives, manifesting related negative effects and social harms. Increase in alcohol outlets in the Southern African countries such as South Africa calls for concern as it has become a social and a political issue [47]. Our study considered problem drinking based on many studies reported on alcohol use among high school learners in South Africa [18,21,47]. However, there is limited research in the townships (i.e., per-urban) on the drivers of alcohol use among adolescents in terms of access to alcohol outlets, drinking motives and harmful effects, except for minimal research [17,19,20]. The current study will enable understanding the context in which alcohol is accessed by learners by exploring by concentrating on the accessibility, motivations and effects of alcohol use among high school learners in Tshwane North and West, South Africa.

2. Methods and Materials

2.1. Study Design and Frameworks

A descriptive cross-sectional study using a quantitative method was conducted among high school learners to determine to accessibility, motivations and effects of alcohol use in the context of poor demographics and infrastructure. In conducting this study, the ecological system theory was relevant to study alcohol issues among high school learners in relation to the external environmental factors, such as taverns and bottle stores. The proximity of alcohol outlets to schools predisposes learners to lifestyle behavioural change. The theory further explains how people living in these communities learn behaviours and grow through interaction, inevitably influencing one another in every aspect of life [48]. The application of the theory to the current study shows the influence of the external environment on individual behaviours. Therefore, interacting with alcohol outlets puts learners at more risk of consuming alcohol than those who are less exposed to alcohol outlets [49,50]. In addition, we also integrated the motivational model, postulating that individuals drink to obtain positive outcomes or to avoid negative consequences (i.e., valence), and they are motivated by internal or external rewards (i.e., source). There are four categories of motivations. The first one is drinking to enhance positive mood (i.e., enhancement), followed by drinking to obtain social rewards (i.e., social motive), then drinking to cope with negative emotions (i.e., coping) and finally, drinking to avoid social rejection (i.e., conformity) [51]. Previous research has found that drinking motives are strong predictors of alcohol consumption and are the most proximal antecedents of alcohol use [35].

2.2. Study Setting and Population

The study was conducted in the peri-urban settings situated in the City of Tshwane municipality, in Gauteng Province (i.e., Soshanguve and Winterveldt). Peri-urban areas, also called townships in South Africa, are underdeveloped residential areas reserved for mostly Africans to live, originating from the Apartheid era [52] and are mostly characterized with poor infrastructure and low socioeconomic status [17,53]. There are 22 high schools located in Soshanguve, with 16 high schools in Tshwane North and 6 high schools in Tshwane West, with a total enrolment of approximately 24,000. Schools in these localities depend on funding from the South African government (i.e., quintile three) [54]. Learners attending high schools in the abovementioned areas participated in the study. These learners received parental consent to be part of the study, gave individual written consent (learners aged $\geq$ 18 years) and assent (learners below 18 years old) (i.e., inclusion criteria). The study excluded learners without parental consent for participation. According to the literature, the adolescent population structure can be classified into two groups, which are the early adolescent population aged 10–14 years and the late adolescent population comprising those aged $\geq$ 15 years [55].

2.3. Sample Size and Sampling Techniques

A minimum representative sample size of 379 was calculated using a validated Raosoft® online sample size calculator (Raosoft Inc., Seattle, WA, USA), considering a total population size of ±24,000 learners, a 95% confidence level and 50% response distribution. To cater for a non-response, the sample was buffered with 10% and increased to 416. After 406 adolescents responded (97% response rate), and 3 were discarded with missing data, a final sample size of 403 was obtained. A multistage sampling was used, firstly randomly selecting 3 high schools out of the 23 schools, followed by a random selection of learners from the lists of learners who have obtained parental consent. After obtaining the permission to conduct the study from the schools' governing body, the main researcher visited the selected schools to recruit and engage learners on the procedure and preparations for data collection with the help of the nutrition teachers. All engagements were conducted during breaks and after schools, and classes were not interrupted.

2.4. Data Collection and Tool

Data were collected using a validated researcher-administered questionnaire, adapted from studies conducted among high school learners [17,19,56] and on motivations for alcohol use [34,36]. Data collected included the demographics, accessibility, motivations and effects of alcohol use among learners. The questionnaire was validated for content, representing all aspects of the constructs on accessibility, motivations and effects of alcohol use, in addition to face validity, which presented all variables to be measured at face value. Following the approval of experts on the suitability of the questionnaire, a mini-study was conducted among 30 learners who did not form part of the main study to pre-test the tool (i.e., questionnaire) for feasibility. During a pilot study, the research assistants who speak the local language (i.e., Setswana) were taken through the process of conducting preliminary interviews for training purposes by the main researcher. After pretesting the questionnaire, slight changes were implemented on wording and layout of the questionnaire.

2.5. Data Analysis

Data were analysed using STATA 17 (StataCorp. 2021. Stata Statistical Software: Release 17., StataCorp LLC, College Station, TX, USA). Missing data were identified through complete case analysis, and three questionnaires with $\geq$10% of missing data were discarded. Descriptive statistics in the form of frequencies (n) and percentages (%) were used to summarise all the categorical variables for accessibility, motivations and effects of alcohol use, and further analyses on comparison by groups were computed using Chi-square (χ^2) and Fishers' exact for variables with expected values less than five in a cell. Significance level was set at $p < 0.05$.

2.6. Ethical Considerations

Ethical approval was obtained from the Sefako Makgatho Health Sciences University Research and Ethics Committee (SMUREC). The project was approved on 1 March 2018, with the projection identification quote of SMUREC/H/51/2018: PG. The study received permission from the Tshwane Department of Education (Ref 8/4/4/1/2). Written informed consent was obtained from parents of all learners and from learners aged 18 years and above, while leaners below 18 years gave assent.

3. Results

Demographic Characteristics of School Learners

The demographics of the school learners are summarised in Table 1. A total sample of 403 learners was obtained from the three selected high schools. The study entailed 185 (46%) boys and 218 (54%) girls, with a mean age of 16 years (SD = ±2), ranging from 10 to 21 years. Two age groups were created; 10–14 years [n = 100 (25%)] and 15–21 years [n = 303 (75%)]. Learners were distributed across the school grades 8 to 12 and characterised by repeating a grade (33%) and missing school at least three times in a quarter of a year, mainly due alcohol hangover (21%) and illnesses (70%). Most learners walked (61%) to school and were receiving pocket money of at least ZAR 100/week (83%). Alcohol use among learners was estimated at 48%, with beer/cider (55%) and wine (27%) being commonly consumed alcohol types. Approximately 18% of learners were smoking, and among the smokers, cannabis was commonly use (62%) compared to cigarette (38%). The demographic differences between boys and girls were significant for age ($p = 0.003$), school grade ($p = 0.048$), repeated grade ($p < 0.001$), transport mode to school ($p = 0.006$) and receiving pocket money ($p = 0.038$).

Table 1. Demographic characteristics of school learners.

Variables	All	Boys	Girls	*p*-Value
	n (%)	n (%)	n (%)	
Age (years)				
10–14	100 (25)	33 (18)	67 (31)	0.003 *
≥15	303 (75)	152 (82)	151 (69)	
School grade				
8	92 (23)	31 (17)	61 (28)	0.048 *
9	90 (22)	41 (22)	49 (22)	
10	57 (14)	32 (17)	25 (11)	
11	91 (23)	42 (23)	49 (23)	
12	73 (18)	39 (21)	34 (16)	
Repeated grade				
No	272 (67)	104 (56)	168 (77)	<0.001 *
Yes	131 (33)	81 (44)	50 (23)	
Missing class in a quarter				
No	330 (82)	149 (81)	181 (83)	0.518
Yes	73 (18)	36 (19)	37 (17)	
Reasons for missed class (n = 73)				
Hangover	15 (21)	6 (24)	9 (19)	0.673 ″
Illnesses	51 (70)	16 (64)	35 (73)	
Other	7 (9)	3 (12)	4 (8)	
In a relationship				
No	135 (33)	58 (31)	77 (35)	0.400
Yes	268 (67)	127 (69)	141 (65)	
Partner older (n = 268)				
No	146 (54)	66 (54)	80 (55)	0.804
Yes	122 (46)	57 (46)	65 (45)	
Partner ages (years, n = 122)				
15–19	71 (58)	31 (54)	40 (62)	0.424
20–29	51 (42)	26 (46)	25 (38)	
Transport mode to school				
Walk distance	249 (61)	128 (69)	121 (56)	0.006 *″
Bus	71 (18)	25 (14)	46 (21)	
Taxi	77 (19)	32 (17)	45 (20)	
Parents' car	6 (2)	0	6 (3)	
Receiving pocket money/week				
No	68 (17)	39 (21)	29 (13)	0.038 *
Yes	335 (83)	146 (79)	189 (87)	
Spare time job				
No	365 (91)	162 (88)	203 (93)	0.057
Yes	38 (9)	23 (12)	15 (7)	
Dwelling place				
Brick house	284 (70)	136 (73)	148 (68)	0.454
RDP house	72 (18)	29 (16)	43 (20)	
Shack	47 (12)	20 (11)	27 (12)	
Refrigerator use				
No	17 (4)	10 (5)	7 (3) ‡	0.279
Yes	385 (96)	175 (95)	210 (97)	
Religion				
Non-Christian	305 (76)	53 (29)	45 (21)	0.062
Christian	68 (24)	132 (71)	173 (79)	

Table 1. *Cont.*

Variables	All	Boys	Girls	*p*-Value
	n (%)	n (%)	n (%)	
Current alcohol use				
No	210 (52)	101 (55)	109 (50)	0.358
Yes	193 (48)	84 (45)	109 (50)	
Types of alcohol (n = 193)				
Beer/cider	106 (55)	56 (55)	50 (55)	0.363 ″
Wine	53 (27)	25 (25)	28 (31)	
Spirits	19 (10)	10 (10)	9 (10)	
Traditional beer	15 (8)	11 (10)	4 (4)	
Smoking				
No	160 (82)	86 (92)	74 (80)	0.478
Yes	34 (18)	16 (8)	18 (20)	
Tobacco types (n = 34)				
Cigarette	13 (38)	6 (38)	7 (39)	0.934
Cannabis	21 (62)	10 (62)	11 (61)	

* indicates significant difference ($p < 0.05$); ″ indicates Fischer's exact test used for variables with expected values less than five in a cell; ‡ indicates n = 217 for girls, and RDP stands for Reconstruction Development Programme.

In Figure 1, most learners (68%) reported that they purchase alcohol at the taverns, while 30% at bottle stores and 2% at the supermarkets.

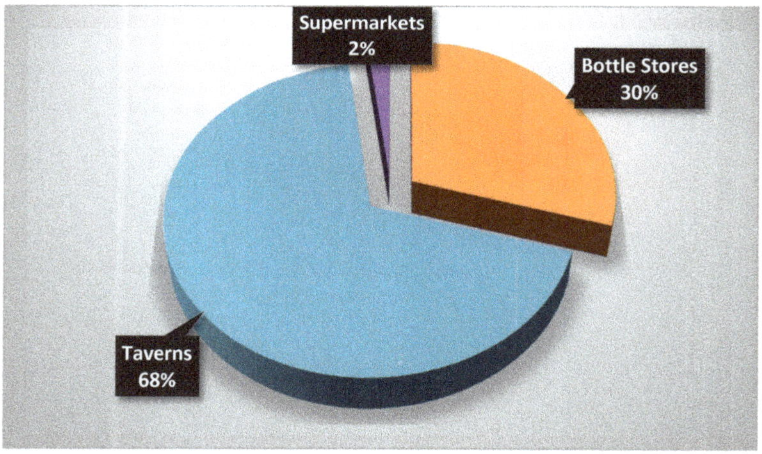

Figure 1. Alcohol outlets for accessibility.

Table 2 shows the accessibility and motivations of alcohol use among school learners, compared using a Chi-squared/Fischer's exact test. Some of the learners (70%) reported that it is wrong to drink alcohol. Over half of the learners (54%) reported that there was a liquor store near their schools, while 70% indicated that they are able to purchase alcohol without using an identity document. Most learners consumed alcohol at parties (41%), a friend's place (22%) and clubs (20%). The last places where learners drank alcohol prior to the study were in the community (40%), at home (25%) and at a social/traditional gathering (25%). Almost half of the learners (48%) introduced themselves to alcohol use, influences of friends (36%) and family (14%). Two-thirds of the learners (70%) reported having friends who consume alcohol, while 79% reported that their family members were drinking alcohol regularly, including siblings (26%) and extended family (42%).

Table 2. Accessibility and motivations of alcohol use among school learners.

Alcohol Access and Use	All	Boys	Girls	p-Value
	n (%)	n (%)	n (%)	
Is it wrong for learners to drink alcohol?				
No	60 (15)	28 (15)	32 (15)	
Not sure	61 (15)	26 (14)	35 (16)	0.855
Yes	282 (70)	131 (71)	151 (69)	
Liquor store near the schools				
No	185 (46)	94 (51)	91 (42)	0.069
Yes	218 (54)	91 (49)	127 (58)	
Identity document required				
No	135 (70)	73 (72)	62 (68)	
Yes	17 (9)	7 (7)	10 (11)	0.600
Sometimes	41 (21)	22 (21)	19 (21)	
Places that learners drink alcohol at (n = 193)				
Party	78 (41)	34 (33)	44 (48)	
Friend's place	43 (22)	26 (25)	17 (19)	
Clubs	38 (20)	25 (25)	13 (14)	0.008 *"
Bottle store	18 (9)	5 (5)	13 (14)	
Traditional rituals	16 (8)	12 (12)	4 (8)	
Last place drank alcohol (n = 193)				
Community site	76 (40)	44 (43)	32 (36)	
Home site	49 (25)	26 (25)	23 (25)	
Social/traditional gatherings	49 (25)	25 (25)	24 (26)	0.501 "
I do not remember the site	18 (9)	7 (7)	11 (12)	
School site	1 (1)	0	1 (1)	
Who introduced you to alcohol (n = 193)				
Self	94 (48)	49 (48)	45 (50)	
Friend	69 (36)	34 (33)	35 (38)	0.420 "
Family	27 (14)	18 (18)	9 (10)	
Neighbour	3 (2)	1 (1)	2 (2)	
Do you have friends who drink alcohol?				
No	119 (30)	58 (31)	61 (28)	0.460
Yes	284 (70)	128 (69)	157 (72)	
Do you have family member who drink alcohol?				
No	86 (21)	45 (24)	41 (19)	0.178
Yes	317 (79)	140 (76)	177 (81)	
Family member drinking alcohol				
Father	60 (19)	30 (21)	30 (17)	
Mother	15 (5)	9 (6)	6 (4)	
Both parents	29 (9)	16 (12)	13 (7)	0.250
Siblings	81 (26)	33 (24)	48 (27)	
Extended Family	132 (42)	52 (37)	80 (45)	

* indicates significant difference ($p < 0.05$); and " indicates Fischer's exact test used for variables with expected values less than five in a cell.

In Table 3, the motivations for alcohol use are reported and compared by sex using Chi-square/Fischer's exact tests. School learners reported that the main motivation for alcohol use was self-pleasure (36%), coping with stress (23%) and increasing their self-esteem (19%), while other learners said it is good for health (10%). Very few indicated that they drink alcohol for fun (8%), or it helps them to function (3%), with insignificant differences by sex for the trend ($p = 0.278$) and between categories.

Table 3. Motivations for alcohol use among the school learners, n = 194.

Proportions of Alcohol Use and Reasons	All N = 403 n (%)	Boys N = 185 n (%)	Girls N = 218 n (%)	*p* Value
Non alcohol users				0.358
No	210 (52)	101(55)	109 (50)	
Yes	193 (48)	84 (45)	109 (50)	0.358
Total	403 (100)	185 (100)	218 (100)	
Alcohol use reasons (n = 193)				0.278 "
Self-pleasure	70 (36)	35 (34)	35 (38)	0.369
It helps me to cope with stress	46 (24)	25 (24)	21 (23)	0.363
It increases my self esteem	38 (19)	20 (20)	18 (20)	0.450
It is good for my health	19 (10)	12 (12)	7 (8)	0.643
I drink alcohol for fun	15 (8)	8 (8)	7 (8)	0.602
It helps me to function	5 (3)	2 (2)	3 (3)	1.000 "
Total	193 (100)	102 (100)	91 (100)	

" indicates Fischer's exact test used for variables with expected values less than five in a cell.

Table 4 shows the effects of alcohol use among school learners compared by sex using Chi-square and Fischer's exact tests. A total of 83% of the learners reported to have knowledge on the effects of alcohol. Most learners believed that alcohol has a negative impact on health (56%), while 25% reported the effect on brain function, 12% on school work and few felt that there is not much effect (7%). Feelings of sadness (45%) and embarrassment (40%) were reported among learners, especially after drinking alcohol. Alcohol-related social harms affecting learners ranged from problems with friends (25%) and parents (17%), physical fights (19%), risky sexual behaviour (11%), injuries/accidents (10%), violence (10%) and engaging in other substance abuse (7%), with insignificant differences by sex for the trend ($p = 0.319$) and between categories.

Table 4. Effects of alcohol use among school learners.

Effects of Drinking Alcohol	All N = 403 n (%)	Boys N = 185 n (%)	Girls N = 218 n (%)	*p*-Value
Knowledge of the effects of alcohol?				0.378
No	69 (17)	35 (19)	34 (16)	
Yes	334 (83)	150 (81)	184 (84)	
Effects of alcohol use				0.143
Negatively impact health	186 (56)	77 (52)	109 (60)	
Brain functioning	84 (25)	36 (24)	48 (26)	
School work	40 (12)	23 (15)	17 (9)	
Has no effect	22 (7)	13 (9)	9 (5)	
Feeling after drinking alcohol				0.079
Sad	59 (45)	19 (35)	40 (52)	
Embarrassed	53 (40)	28 (52)	25 (32)	
Just fine	19 (15)	7 (13)	12 (16)	
Alcohol related harm in the past 12 months (n = 193)				0.319
Problem with friends	48 (25)	27 (26)	21 (23)	0.586
Physical fight	37 (19)	16 (15)	21 (23)	0.193
Problem with parents	33 (17)	13 (13)	20 (22)	0.089
Engaging in unprotected sex	21 (11)	12 (13)	9 (11)	0.651
Injury/accident	20 (11)	14 (13)	6 (7)	0.189
Trouble with police officers	20 (10)	12 (12)	8 (8)	0.499
Engaged in other substances	14 (7)	8 (8)	6 (6)	0.738
Total	193 (100)	102 (100)	91 (100)	

4. Discussion

We investigated the accessibility, motivations and effects of alcohol use among high school learners. Poor demographic and lifestyle characteristics were observed among learners in terms of alcohol and tobacco use at an early age. These included having relationships with older partners, walking a distance to school, having jobs during spare time, repeating grades and missing school mainly due alcohol hangover and illnesses. Peri-urban areas, also called townships in South Africa, are known for several challenges in terms of lack of social and economic facilities required to build sustainable communities [53], with low socioeconomic status [17]. Both poor socioeconomic status and infrastructure have been associated with increased alcohol use, promoted by accessibility and availability of alcohol products in the neighbourhood [56–58]. This is in addition to poverty and social inequality prevalent in South Africa and implicated in alcohol use, especially heavy drinking [31]. The wide availability of alcohol has been related to poor education performance and higher rates of school absenteeism among young people [41].

Accessibility to alcohol outlets close to schools was reported by learners in this study. Further results showed that learners were purchasing alcohol from taverns and bottle stores, and two-thirds reported purchasing without an identity document required. Despite the fact that most of the learners reported that alcohol use is wrong, almost half of them were using alcohol. Learners also consumed alcohol at parties, friend's places, clubs and communities. These findings confirm ease of alcohol access by learners both around schools and in the communities, consistent with studies conducted in South Africa [21,24,31] and other countries [39,59,60]. Hasheena et al. [28], reported that 62% of grade 8 learners, 81% of grade 10 learners and 92% of grade 12 learners reported that alcohol was easily accessible to them in Namibia. Studies have reported that the more alcohol outlets within school vicinity and in the communities, the more likely youths will engage in drinking alcohol, and this has been reported in countries such as South Africa [21,24,31], Namibia [39], Brazil [59] and Canada [60]. Therefore, the wide availability of alcohol outlets around schools and the lack of regulating policies contribute to the normalisation of alcohol use [60–63]. In the South African context, the National Liquor Policy prohibits alcohol outlets operating within a 500 m radius to schools and other public places, such as churches, suggesting alcohol outlets are violating the Liquor Policy [23]. Recent literature suggests that South Africa does not have national restrictions on the density of alcohol outlets or the days, hours and location of their operation as of mid-2018 [64]. The more that alcohol outlets in the community operate without age restriction entry, the younger the people who are exposed to drinking alcohol, and that has fatal health outcomes.

Motivations for alcohol use among learners in the current study were self-pleasure, increasing self-esteem and coping with stress. Abbey et al. [65] have reported that a belief that alcohol helps with coping or social skills puts individuals at risk for abusing alcohol, especially when the appropriate environmental circumstances arise. Studies show that people who reported a higher motivation to drink alcohol in order to cope with negative effects, tend to drink more alcohol, yet encounter more negative alcohol-related consequences [65–67]. Most learners, also indicated that after consuming alcohol, they felt sad and embarrassed. Additional studies have found that students reported sadness, fear and shyness after alcohol use [68]. It seems as if alcohol use in the current study was motivated by learners' feelings, and seeking coping mechanisms, without considering the emotional impact that accompanies alcohol use. Sadness was the most identified feeling among learners in this study, especially after drinking alcohol, which is similar to other researchers' suggestion [69].

Most of the learners in this study indicated that they introduced themselves to alcohol use, although friends and families were also implicated. The current findings correspond with the findings from other studies in South Africa [17,37] that showed that parents who use alcohol have great influence on their adolescents' drinking behaviour, and the curiosity of experimenting with alcohol use is peer related [70,71]. Further literature documents that some of the strongest influences on adolescent drinking behaviour come from the people that adolescents spend most time with, such as family and friends [38,72,73]. Although the influence of parents on alcohol use among their children diminished with time, parents seem to exert a greater influence among their children before they reach the age of 15 [74]. Social learning has been implicated as an influence on adolescent alcohol use based on the behaviour of other significant persons such as family members and friends, as well as social imitation and social modelling from peers. as reported by Bențea [36]. Parents' motivation, norms, beliefs, values and goals are imperative if they are to modify behaviour in their adolescent children. Unquestionably, people shift from the social context of the family unit during childhood to focus more on their peers and their schools during adolescence, which might explain the situation in the current study. Nonetheless, parents must promote, monitor and manage the behaviour of adolescents.

Knowing the effects of alcohol use was common among the learners in the current study. They reported that alcohol use has a negative impact on health, brain function and school work. The alcohol harm paradox suggests that youth living in lower income areas are more exposed to alcohol and suffer more alcohol-related harms than their peers living in less deprived neighbourhoods [61,62]. Early initiation of alcohol use is well documented to lead to alcohol use disorder in adolescence and later in adulthood [41,59]. Behavioural research has found that educational performance is highly correlated with substance abuse [73], and a high level of drinking results in self-reported academic difficulty among high school learners [74]. Additionally, alcohol has toxic effects predisposing users to chronic diseases [26,33,75], especially later in adulthood [41,42]. Learners further reported that in the past 12 months, due to alcohol use, they encountered social harms with friends and parents, had physical fights, injury/accidents, engaged in unprotected sex, had trouble with police and engaged in other substance abuse. Research on alcohol use and delinquent behaviour among adolescents in South Africa has reported that 12.4% were found carrying a handgun in the previous year, while 22.4% had attacked someone with the intention of inflicting serious harm, and 41.3% were involved in serious fights at school [71]. Further, Windle and Zucker [76] have reported that adolescents who regularly consume alcohol are more likely to indulge in other risky behaviours, including smoking, recreational drug use and risky sexual behaviour. Other researchers in South Africa have associated harmful alcohol use with increased incidence of risky sexual risk behaviour and consequently an increased HIV risk in youth [46], negatively impacting the public health in South Africa.

5. Limitations

In terms of limitations, the study may not necessarily represent the entire population of adolescent learners, since only black school learners from the peri-urban high schools participated in the study. This has implications for the generalisability of the findings to other races and settings. Causality could not be concluded from the results due to the cross-sectional and descriptive nature of the study, respectively. Relying exclusively on adolescent self-reports, measures of alcohol consumption might be affected by social desirability, even if anonymity and confidentiality were assured. Lastly, we did not measure biochemical variables that confirm alcohol use, such as plasma activities of gamma-glutamyl transpeptidase, aspartate aminotransferase and the erythrocyte mean corpuscular volume. Future research should endeavour to use the Alcohol Use Disorders Identification Test to screen for alcohol use and study plasma activities of the abovementioned variables and several alcohol-related clinical outcomes. The conclusion of this study could only provide a starting point for other long-term prospective research focusing on identifying and understanding the ease of access to alcohol outlets with motivations and social harms, which might explain alcohol use through the phases of adolescence.

6. Conclusions

The availability of alcohol outlets (i.e., taverns and bottle stores) around schools and easy access to alcohol without enforcing age restrictions support adolescent alcohol use even though the National Liquor Act in South Africa prohibits alcohol sales to individuals younger than 18. There is a need to prioritize limiting adolescent alcohol access and enforcing the South African National Liquor Act. Although alcohol-related problems in school adolescents have been attributed to urban environments, this study confirms that peri-urban settings (i.e., townships) have as much influence. Continuous research on neighbourhood characteristics of alcohol use among school age adolescents is useful for planning public health programs. Additionally, life skills training to address learners' drinking motives and constantly alerting parents about the relevance of modelling behaviour should be emphasised.

Author Contributions: Conceptualization, resources, software, formal analysis, and original draft, M.M.; software, formal analysis, and original draft., L.C.; software, formal analysis, validation, original draft, and review draft, P.M. All authors have read and agreed to the published version of the manuscript.

Funding: This research received no external funding.

Institutional Review Board Statement: This study received ethical clearance from Sefako Makgatho Health Sciences University Research and Ethics Committee (SMUREC), on 1 March 2018, with identification code of SMUREC/H/51/2018: PG.

Informed Consent Statement: Informed consent was obtained from the parents of the learners, as well as from learners aged 18 years and above, while assent was obtained from learners below 18 years.

Data Availability Statement: The dataset for the school-age adolescents generated and analysed during the current study is available from the corresponding author upon reasonable request.

Acknowledgments: The authors thank the learners and their parents in the high schools of Soshanguve and Winterveldt townships for giving consent and assent for participation in the study, as well as the Tshwane Department of Education for giving us permission to conduct the study, and the research assistants for assisting with the study.

Conflicts of Interest: The authors declare no conflict of interest.

References

1. Onaivi, E. Drug Addiction—A Global Problem for the Rich and Poor. *Trop. J. Pharm. Res.* **2009**, *8*, 191–192.
2. WHO. *Global Status Report on Alcohol and Health*; World Health Organization: Geneva, Switzerland, 2011. Available online: http://apps.who.int/iris/bitstream/handle/10665/44499/9789241564151_eng.pdf;jsessionid=59D816495A9F80BAF76CD79F01CE24D4?sequence=1 (accessed on 22 May 2022).
3. Jumbe, S.; Kamninga, T.M.; Mwalwimba, I.; Kalu, U.-G. Determinants of adolescent substance use in Africa: A systematic review and meta-analysis protocol. *Syst. Rev.* **2021**, *10*, 125. [CrossRef]
4. Ogundipe, O.; Amoo, E.; Adeloye, D.; Olawole-Isaac, A. Substance use among adolescents in sub-Saharan Africa: A systematic review and meta-analysis. *S. Afr. J. Child Health* **2018**, *2018*, s79–s84.
5. Matshipi, M.; Monyeki, K.D.; Mafumo, N.; Monyeki, S.M.; Siweya, H.J.; Kemper, H.C.G. The Use of Alcohol and Knowledge of Cardiovascular Diseases among Ellisras Rural Children Aged 14–22 Years: Ellisras Longitudinal Study. *Int. J. Environ. Res. Public Health* **2019**, *16*, 2650. [CrossRef] [PubMed]
6. Modjadji, P.; Pitso, M. Maternal Tobacco and Alcohol Use in Relation to Child Malnutrition in Gauteng, South Africa: A Retrospective Analysis. *Children* **2021**, *8*, 133. [CrossRef]
7. Rantao, M.; Ayo-Yusuf, O.A. Dual use of cigarettes and smokeless tobacco among South African adolescents. *Am. J. Health Behav.* **2012**, *36*, 124–133. [CrossRef]
8. Reddy, P.S.; James, S.; Resnicow, K.; Sewpaul, R.; Masuka, P.; van den Borne, B. Prevalence and Correlates of Smokeless Tobacco Use Among Grade 8–11 School Students in South Africa A Nationwide Study. *Nicotine Tob. Res.* **2014**, *16*, 1167–1173. [CrossRef] [PubMed]
9. Mashita, R.J.; Themane, M.J.; Monyeki, K.D.; Kemper, H.C.G. Current smoking behaviour among rural South African children: Ellisras Longitudinal Study. *BMC Pediatr.* **2011**, *11*, 58. [CrossRef]
10. Monyeki, K.D.; Kemper, H.C.G.; Amusa, L.O.; Motshwane, M. Advertisement and knowledge of tobacco products among Ellisras rural children aged 11 to 18 years: Ellisras Longitudinal study. *BMC Pediatr.* **2013**, *13*, 111. [CrossRef]
11. Carney, T.; Myers, B.J.; Louw, J.; Lombard, C.; Flisher, A.J. The relationship between substance use and delinquency among high-school students in Cape Town, South Africa. *J. Adolesc.* **2013**, *36*, 447–455. [CrossRef]

12. Kaufman, Z.A.; Braunschweig, E.N.; Feeney, J.; Dringus, S.; Weiss, H.; Delany-Moretlwe, S.; Ross, D.A. Sexual risk behavior, alcohol use, and social media use among secondary school students in informal settlements in Cape Town and Port Elizabeth, South Africa. *AIDS Behav.* **2014**, *18*, 1661–1674. [CrossRef] [PubMed]
13. Parry, C.D.; Myers, B.; Morojele, N.K.; Flisher, A.J.; Bhana, A.; Donson, H.; Plüddemann, A. Trends in adolescent alcohol and other drug use: Findings from three sentinel sites in South Africa (1997–2001). *J. Adolesc.* **2004**, *27*, 429–440. [CrossRef]
14. Reddy, S.P.; James, S.; Sewpaul, R.; Koopman, F.; Funani, N.I.; Sifunda, S.; Josie, J.; Masuka, P.; Kambaran, N.S.; Omardien, R.G. *Umthente Uhlaba Usamila—The 2nd South African Youth Risk Behaviour Survey 2008*; DoH: Pretoria, South Africa, 2010. Available online: http://www.mrc.ac.za/healthpromotion/yrbs_2008_final_report.pdf (accessed on 28 May 2022).
15. Reddy, S.; James, S.; Sewpaul, R.; Sifunda, S.; Ellahebokus, A.; Kambaran, N.S.; Omardien, R.G. *Umthente Uhlaba Usamila: The 3rd South African National Youth Risk Behaviour Survey 2011*; South African Medical Research Council: Cape Town, South Africa, 2013.
16. Reddy, S.P.; Panday, S.; Swart, D.; Jinabhai, C.C.; Amosun, S.L.; James, S.; Monyeki, K.D.; Stevens, G.; Morejele, N.; Kambaran, N.S.; et al. *Umthenthe Uhlaba Usamila—The 1st South African Youth Risk Behaviour Survey 2002*; DoH: Pretoria, South Africa, 2022.
17. Chauke, T.M.; van der Heever, H.; Hoque, M.E. Alcohol use amongst learners in rural high school in South Africa. *Afr. J. Prim. Health Care Fam. Med.* **2015**, *7*, a755. [CrossRef] [PubMed]
18. Maserumule, O.M.; Skaal, L.; Sithole, S.L. Alcohol use among high school learners in rural areas of Limpopo province. *S. Afr. J. Psychiatry* **2019**, *25*, a1183. [CrossRef] [PubMed]
19. Mohale, D.; Mokwena, K.E. Substance use amongst high school learners in the south of Johannesburg: Is this the new norm? *S. Afr. Fam. Pract.* **2020**, *62*, a5122. [CrossRef]
20. Mokwena, K.; Sindane, P. Cigarette smoking and alcohol intake among high school learners in Pretoria, South Africa. *Psychol. Educ. J.* **2020**, *57*, 531–536.
21. Onya, H.; Tessera, A.; Myers, B.; Flisher, A. Adolescent alcohol use in rural South African high schools: Original. *Afr. J. Psychiatry* **2012**, *15*, 352–357. [CrossRef]
22. Wagenaar, A.C.; Toomey, T.L.; Erickson, D.J. Preventing youth access to alcohol: Outcomes from a multi-community time-series trial. *Addiction* **2005**, *100*, 335–345. [CrossRef]
23. Department of TradeandIndustry. Final Liquor Policy Paper National Liquor Policy Review. 2016. Available online: https://www.gov.za/sites/default/files/gcis_document/201609/40321gon1208.pdf (accessed on 1 June 2022).
24. Letsela, L.; Weiner, R.; Gafos, M.; Fritz, K. Alcohol Availability, Marketing, and Sexual Health Risk Amongst Urban and Rural Youth in South Africa. *AIDS Behav.* **2019**, *23*, 175–189. [CrossRef]
25. Babor, T.F.; Robaina, K.; Jernigan, D. The influence of industry actions on the availability of alcoholic beverages in the African region. *Addiction* **2015**, *110*, 561–571. [CrossRef]
26. Babor, T.F.; Babor, T.; Caetano, R.; Casswell, S.; Edwards, G.; Giesbrecht, N.; Grube, J.W.; Graham, K. *Alcohol: No Ordinary Commodity: Research and Public Policy*; Oxford University Press: Oxford, UK, 2010.
27. Gorman, D.M.; Speer, P.W.; Gruenewald, P.J.; Labouvie, E.W. Spatial dynamics of alcohol availability, neighborhood structure and violent crime. *J. Stud. Alcohol* **2001**, *62*, 628–636. [CrossRef] [PubMed]
28. Hasheela, M.W.; Makhubele, J.C.; Ananias, J.A.; Matlakala, F.K.; Mafa, P.; Chiwalo, B.N.; Rapholo, S.F.; Freeman, R.J.; Svinurai, A.; Hamuse, T.N.I. Proximity and Density of Alcohol outlets as a Risk Factor of Alcohol Abuse amongst the Youth: A Case Study of Border Town in Northern Region of Namibia. *Glob. J. Health Sci.* **2019**, *11*, 134–141. [CrossRef]
29. Hay, G.C.; Whigham, P.A.; Kypri, K.; Langley, J.D. Neighbourhood deprivation and access to alcohol outlets: A national study. *Health Place* **2009**, *15*, 1086–1093. [CrossRef]
30. Lebese, R.; Ramakuela, N.; Maputle, S. Perceptions of teenagers about substance abuse in Muyexe Village, Mopani District of Limpopo Province, South Africa. *Afr. J. Phys. Health Educ. Recreat. Danc.* **2014**, *20*, 329–347.
31. Rabotata, E.; Makhubele, J.; Mafa, P. Accessibility of alcohol as a risk factor contributing towards alcohol abuse amongst the youth of a selected rural community in Limpopo Province, South Africa. *Tech. Soc. Sci. J.* **2021**, *25*, 715–721. [CrossRef]
32. Kuther, T.L.; Timoshin, A. A comparison of social cognitive and psychosocial predictors of alcohol use by college students. *J. Coll. Stud. Dev.* **2003**, *44*, 143–154. [CrossRef]
33. WHO. *Global Status Report on Alcohol and Health 2018*; World Health Organization: Geneva, Switzerland, 2019.
34. Bergagna, E.; Tartaglia, S. Drinking Motives, Perceived Norms, and Adolescents' Drinking. *J. Drug Issues* **2019**, *49*, 3–14. [CrossRef]
35. Tartaglia, S.; Fedi, A.; Miglietta, A. Family or friends: What counts more for drinking behaviour of young adults?/Familia o amigos:¿qué pesa más en los hábitos de consumo de alcohol de los jóvenes? *Rev. Psicol. Soc.* **2017**, *32*, 1–22. [CrossRef]
36. Benţea, C.-C. Motivations for Alcohol Use in Late Adolescence and Educational Strategies of Intervention. *Procedia-Soc. Behav. Sci.* **2014**, *128*, 186–191. [CrossRef]
37. Morojele, N.; Parry, C.; Brook, J. Substance abuse and the young: Taking action. *MRC Res. Brief* **2009**, *6*, 1–4.
38. Walsh, S.D.; Djalovski, A.; Boniel-Nissim, M.; Harel-Fisch, Y. Parental, peer and school experiences as predictors of alcohol drinking among first and second generation immigrant adolescents in Israel. *Drug Alcohol Depend.* **2014**, *138*, 39–47. [CrossRef] [PubMed]
39. Frank, R.; Makhubele, J.; Ananias, J.; Svinurai, A.; Winnie, M.; Tiberia, H.; Ilonga, N.; Freeman, R.; Mafa, P.; Matlakala, F.; et al. Non-Governmental Organisations Personnel, Social Workers and Religious Leaders' Perspectives on the Risk Factors of Alcohol Abuse Amongst the Youth: The Case of the Northern Region of the Republic of Namibia. *Glob. J. Health Sci.* **2019**, *11*, 55–64. [CrossRef]

40. Wilson, A.; Stearne, A.; Gray, D.; Saggers, S. *The Harmful Use of Alcohol amongst Indigenous Australians*; Australian Indigenous Health Info Net: Mount Lawley, Australia, 2010.
41. Newbury-Birch, D. Impact of Alcohol Consumption on Young People: A Review of Reviews. Department for Children, Schools and Families, © Newcastle University 2009. Available online: https://dera.ioe.ac.uk/11355/1/DCSF-RR067.pdf (accessed on 22 May 2022).
42. Ystrom, E.; Kendler, K.S.; Reichborn-Kjennerud, T. Early age of alcohol initiation is not the cause of alcohol use disorders in adulthood, but is a major indicator of genetic risk. A population-based twin study. *Addiction* **2014**, *109*, 1824–1832. [CrossRef] [PubMed]
43. Whiteford, H.A.; Degenhardt, L.; Rehm, J.; Baxter, A.J.; Ferrari, A.J.; Erskine, H.E.; Charlson, F.J.; Norman, R.E.; Flaxman, A.D.; Johns, N. Global burden of disease attributable to mental and substance use disorders: Findings from the Global Burden of Disease Study 2010. *Lancet* **2013**, *382*, 1575–1586. [CrossRef]
44. Kalichman, S.C.; Simbayi, L.C.; Kaufman, M.; Cain, D.; Jooste, S. Alcohol use and sexual risks for HIV/AIDS in sub-Saharan Africa: Systematic review of empirical findings. *Prev. Sci.* **2007**, *8*, 141–151. [CrossRef]
45. Parry, C.D. South Africa: Alcohol today. *Addiction* **2005**, *100*, 426–429. [CrossRef]
46. Morojele, N.K.; Nkosi, S.; Kekwaletswe, C.T.; Saban, A.; Parry, C. *Review of Research on Alcohol and HIV in sub-Saharan Africa*; South African Medical Research Council Policy Brief: Cape Town, South Africa, 2013.
47. Livingston, M. The social gradient of alcohol availability in Victoria, Australia. *Aust. N. Z. J. Public Health* **2012**, *36*, 41–47. [CrossRef] [PubMed]
48. Bronfenbrenner, U. *The Ecology of Cognitive Development: Research Models and Fugitive Findings*; Wozniak, R.H., Fischer, K.W., Eds.; Psychology Press: New York, NY, USA, 2014.
49. Ahmed, S.R.; Amer, M.M.; Killawi, A. The ecosystems perspective in social work: Implications for culturally competent practice with American Muslims. *J. Relig. Spiritual. Soc. Work. Soc. Thought* **2017**, *36*, 48–72. [CrossRef]
50. Roundy, L. Bronfenbrenner's Ecological Systems Theory of Development: Definition & Examples. Retrieved November 2015, from Study. 2006. Available online: http://study.com/academy/lesson/bronfenbrenners-ecological-systems-theory-of-development-definition-examples.html (accessed on 15 June 2020).
51. Cox, W.M.; Klinger, E. A motivational model of alcohol use. *J. Abnorm. Psychol.* **1988**, *97*, 168–180. [CrossRef]
52. Pernegger, L.; Godehart, S. *Townships in the South African Geographic Landscape—Physical and Social Legacies and Challenges*; Training for Township Renewal Initiative: Cape Town, South Africa, 2007.
53. Philip, K. A History of Townships in South Africa. In *Economics of South African Townships: Special Focus on Diepsloot*; World Bank Group: Washington, DC, USA, 2014; pp. 31–49.
54. Dass, S.; Ringuest, A. Basic Education Rights Handbook: Education Rights in South Africa Johannesburg, South Africa: Section 27. 2017. Available online: http://section27orgza/wpcontent/uploads/2017/02 (accessed on 30 May 2022).
55. Makiwane, M.; Chimere-Dan, G.C. *The State of Youth in South Africa: A Demographic Perspective*; Human Sciences Research Council: Pretoria, South Africa, 2009.
56. Morrison, C.; Gruenewald, P.J.; Ponicki, W.R. Race, ethnicity, and exposure to alcohol outlets. *J. Stud. Alcohol Drugs* **2016**, *77*, 68–76. [CrossRef]
57. Badland, H.; Mavoa, S.; Livingston, M.; David, S.; Giles-Corti, B. Testing spatial measures of alcohol outlet density with self-rated health in the Australian context: Implications for policy and practice. *Drug Alcohol Rev.* **2016**, *35*, 298–306. [CrossRef] [PubMed]
58. Myran, D.T.; Chen, J.T.; Bearnot, B.; Ip, M.; Giesbrecht, N.; Rees, V.W. Alcohol availability across neighborhoods in Ontario following alcohol sales deregulation, 2013–2017. *Am. J. Public Health* **2019**, *109*, 899–905. [CrossRef] [PubMed]
59. Martins, J.G.; Guimarães, M.O.; Jorge, K.O.; Silva, C.J.d.P.; Ferreira, R.C.; Pordeus, I.A.; Kawachi, I.; Zarzar, P.M.P.d.A. Binge drinking, alcohol outlet density and associated factors: A multilevel analysis among adolescents in Belo Horizonte, Minas Gerais State, Brazil. *Cad. Saúde Pública* **2019**, *36*, 1–13. [CrossRef] [PubMed]
60. Slaunwhite, A.K.; McEachern, J.; Ronis, S.T.; Peters, P.A. Alcohol distribution reforms and school proximity to liquor sales outlets in New Brunswick. *Can. J. Public Health* **2017**, *108*, e488–e496. [CrossRef]
61. Petruzzi, L.J.; Pullen, S.J.; Lange, B.C.; Parnarouskis, L.; Dominguez, S.; Harris, B.; Quiterio, N.; Lekpeh, G.; Manobah, B.; Henderson, D.C. Contributing risk factors for substance use among youth in postconflict Liberia. *Qual. Health Res.* **2018**, *28*, 1827–1838. [CrossRef]
62. Ibitoye, M.; Kaaya, S.; Parker, R.; Likindikoki, S.; Ngongi, L.; Sommer, M. The influence of alcohol outlet density and advertising on youth drinking in urban Tanzania. *Health Place* **2019**, *58*, 102141. [CrossRef]
63. Yassin, N.; Afifi, R.; Singh, N.; Saad, R.; Ghandour, L. "There is zero regulation on the selling of alcohol": The voice of the youth on the context and determinants of alcohol drinking in Lebanon. *Qual. Health Res.* **2018**, *28*, 733–744. [CrossRef] [PubMed]
64. Parry, C.D.H.; Bennetts, A.L. *Alcohol Policy and Public Health in South Africa*; Oxford University Press: Cape Town, South Africa, 1998.
65. Abbey, A.; Smith, M.J.; Scott, R.O. The relationship between reasons for drinking alcohol and alcohol consumption: An interactional approach. *Addict. Behav.* **1993**, *18*, 659–670. [CrossRef]
66. Cooper, M.L.; Frone, M.R.; Russell, M.; Mudar, P. Drinking to regulate positive and negative emotions: A motivational model of alcohol use. *J. Pers. Soc. Psychol.* **1995**, *69*, 990–1005. [CrossRef]

67. Kuntsche, E.; Stewart, S.H.; Cooper, M.L. How stable is the motive–alcohol use link? A cross-national validation of the Drinking Motives Questionnaire Revised among adolescents from Switzerland, Canada, and the United States. *J. Stud. Alcohol Drugs* **2008**, *69*, 388–396. [CrossRef]
68. Hussong, A.M.; Galloway, C.A.; Feagans, L.A. Coping motives as a moderator of daily mood-drinking covariation. *J. Stud. Alcohol* **2005**, *66*, 344–353. [CrossRef] [PubMed]
69. Gottfredson, N.C.; Hussong, A.M. Parental involvement protects against self-medication behaviors during the high school transition. *Addict. Behav.* **2011**, *36*, 1246–1252. [CrossRef] [PubMed]
70. Hartman, A. *Millennium Life Orientation Grade 9 Learner's Book*; Action Publisher: Paarl, South Africa, 2006.
71. McCance-Katz, E.F. The National Survey on Drug Use and Health: 2017. Substance Abuse and Mental Health Services Administration. Available online: https://www.samhsa.gov/data/sites/default/files/nsduh-ppt-09-2018.pdf (accessed on 7 May 2019).
72. Cruz, J.E.; Emery, R.E.; Turkheimer, E. Peer network drinking predicts increased alcohol use from adolescence to early adulthood after controlling for genetic and shared environmental selection. *Dev. Psychol.* **2012**, *48*, 1390. [CrossRef]
73. Trucco, E.M.; Colder, C.R.; Wieczorek, W.F.; Lengua, L.J.; Hawk, L.W. Early adolescent alcohol use in context: How neighborhoods, parents, and peers impact youth. *Dev. Psychopathol.* **2014**, *26*, 425–436. [CrossRef]
74. Dawson, D.A. The link between family history and early onset alcoholism: Earlier initiation of drinking or more rapid development of dependence? *J. Stud. Alcohol* **2000**, *61*, 637–646. [CrossRef]
75. Rehm, J. How should prevalence of alcohol use disorders be assessed globally? *Int. J. Methods Psychiatr. Res.* **2016**, *25*, 79–85. [CrossRef] [PubMed]
76. Windle, M.; Zucker, R.A. Reducing underage and young adult drinking: How to address critical drinking problems during this developmental period. *Alcohol Res. Health* **2010**, *33*, 29–44. [PubMed]

MDPI AG
Grosspeteranlage 5
4052 Basel
Switzerland
Tel.: +41 61 683 77 34

Children Editorial Office
E-mail: children@mdpi.com
www.mdpi.com/journal/children

www.ingramcontent.com/pod-product-compliance
Lightning Source LLC
Chambersburg PA
CBHW062058220926
51850CB00010B/2810
* 9 7 8 3 7 2 5 8 6 8 3 8 4 *